Higher than Everest

Few challenges remain for Earth-bound adventurers, but do not fear – the Solar System abounds with weird and wonderful places to explore.

In this unique guidebook, Paul Hodge takes us on a tour of the most spectacular sites in the Solar System. His vivid descriptions of the challenges provide a compelling introduction to extra-terrestrial environments. Starting with a climb of Mars' Mt. Olympus, much higher than Everest, you will be taken on imaginary expeditions to such exotic places as the Moon's Alpine Valley, Venus' precipitous and scorching Mt. Maxwell, a table mountain on Io, the snows of Saturn's rings and Miranda's incredibly high, icy cliff. You will be treated to a descent into a fabulous canyon on Mars, one that dwarfs the Earth's Grand Canyon, and will explore the rock lakes and terraces of Copernicus, a giant crater on the Moon.

Who knows – one day these adventures may really take place!

PAUL HODGE is Professor of Astronomy at the University of Washington and Editor in Chief of *Astronomical Journal*. For many years he has taught a popular undergraduate course on the planets, from which this book originated. His research has spanned the range from interplanetary dust to the extragalactic distance scale. Current research includes studies of star-formation and galactic evolution, using the Hubble Space Telescope to investigate nearby galaxies.

Higher than Everest

An Adventurer's Guide to the Solar System

Paul Hodge
University of Washington

CAMBRIDGE
UNIVERSITY PRESS

PUBLISHED BY THE PRESS SYNDICATE OF THE UNIVERSITY OF CAMBRIDGE
The Pitt Building, Trumpington Street, Cambridge, United Kingdom

CAMBRIDGE UNIVERSITY PRESS
The Edinburgh Building, Cambridge CB2 2RU, UK
40 West 20th Street, New York, NY 10011-4211, USA
10 Stamford Road, Oakleigh, VIC 3166, Australia
Ruiz de Alarcón 13, 28014 Madrid, Spain
Dock House, The Waterfront, Cape Town 8001, South Africa

http://www.cambridge.org

First published 2001

Printed in the United Kingdom at the University Press, Cambridge

Typeface Swift-Regular 9.75/14pt. *System* QuarkXpress® [HMCL]

A catalogue record for this book is available from the British Library

Library of Congress Cataloguing in Publications data

Hodge, Paul W.
Higher than Everest: an adventurer's guide to the solar system/Paul Hodge.
    p.   cm.
Includes bibliographical references and index.
ISBN 0 521 65133 6
1. Solar system–Popular works. I. Title
QB501.2.H63 2001
523.4–dc21   00-065145

ISBN 0 521 65133 6 hardback

Contents

Preface

This is a book about the planets and their moons, disguised as a guide-book. It takes you to each planet (and to some really interesting satellites) by describing a possible adventure involving certain spectacular planetary features. Each adventure is hypothetically possible, though none is easy and it will be many years before anyone will be taking any of these trips. In many cases, the extreme conditions to be encountered are truly formidable and they will require remarkable spacecraft and equally remarkable space suits. But I believe that each of them will one day be done. What seemed impossible not long ago is now routine.

Of course, recently we have experienced something of a halt in adventurous space travel. It seems a long time ago since the Apollo program. Apparently, for the moment, humanity is more willing to spend its extra money on other things. Some of these other things are very important. Some of the world needs to expend its resources on its people, who need adequate food and shelter, education and a society that serves its individual members. Other parts of the world seem to have less desperate needs and they are satisfying their appetites for more luxurious homes, bigger vehicles and newer computers. For all of the world there is a critical lack of concern for the Earth's future and more effort and resources must be spent on preserving our planet's health and livability. I believe, though, that the time will come when the world will be able to catch up with the space agencies' marvelous far-seeing vision, which has been kept alive even in periods of budget cuts and political apathy. We will explore the planets and, after the initial scientific work is done, some of us will not be able to resist the call of adventure. Until that time arrives, you can begin planning some spectacular trips ahead of time by reading this guidebook.

Note on technicalities: The descriptions of the planets' and moons' surfaces given in this book are based on current understanding. Details will certainly change as we study them further. I have tried to indicate where scientists are presently uncertain about the details and, of course, the adventures themselves are hypothetical. I am thankful to the scientists who have reviewed the text, especially to Dr. Toby Smith.

Note about the figures: Most of the images of the planets and moons in this book are based on NASA photographs and USGS maps. Some of the NASA images were derived from those provided on the Web by Malin Space Science Systems. Dr. Dale Smith generously provided Figure 6.8, Gordon and Sara Hodge supplied Figure 6.1 and Dr. George Wallerstein provided Figure 15.5.

Higher than Everest

This guidebook begins with some adventurous expeditions on the planet Mars. Though not the closest planet to Earth (Venus is closer), Mars seems the most Earth-like and it has long been the subject of romantic tales of alien beings and exotic lands.

We now know that Martians don't exist, but the exotic landscape of Mars does, beckoning for us to come, to map, and to explore. Mars has many spectacular features and a trip to any of these places on the surface of Mars will be a great adventure.

Our most massive mountain

Your first expedition will be to the highest volcano in the Solar System, the highest mountain known to Man. Mt. Olympus on Mars is far higher and vastly more massive than any mountain on Earth. It is the ultimate challenge of mountain-climbing among the planets. And yet, compared with climbing Mt. Everest or many other terrestrial peaks, reaching its top is not technically difficult. Following the route described here requires no tricky rock climbing, no glacier travel and no roping up.

Of course, for a mountain of this height a climber will need oxygen. Even the strongest and most daring free-climbing mountain-conqueror of the Himalayas needs oxygen tanks on Olympus, just as he or she needs them anywhere on Mars. Even down on the Martian plains the air is thin, less than 1% of the air at the Earth's sea level, and it contains no oxygen at all, mostly just carbon dioxide.

It's also pretty cold on Mars. Temperatures at night on the plains drop to $-200\,°$F and fall even lower on the heights of Olympus. But a standard Martian space suit should suffice for maintaining reasonable comfort for the climb. The hardships and the dangers that are standard for the Earth's famous climbs are much worse than what a well-equipped climbing expedition to Olympus can expect. The mountain is huge, but its slopes are not steep, except in a few, rather surprising places. For the most part it is a gentle giant.

Mt. Olympus is located near the tropics of Mars, at a latitude of $18\,°$ north, similar to Earth's Hawaiian Islands. But even its base never gets

The Tharsis region of Mars. In this map, based on elevations measured by the Mars Global Surveyor, elevations are color-coded. The Tharsis Dome is red and yellow and the high volcanoes are white. Olympus is the peak at the left of the bulge. Mariner Valley extends across the bottom of the picture to the right (NASA).

warm, as the mountain rises from the edge of a high, cold upland area called the Tharsis Dome. This huge, volcano-rich plain averages a height of some 10,000 feet above Martian "sea level," a reference level roughly equivalent to Earth's sea level. The Tharsis Dome is immense, spanning some 4,000 miles. On Earth it would cover most of North America. Its round dome shape is the apparent result of concentrated up-welling of molten rock from Mars' interior, and the dome is punctuated with giant volcanoes of various ages, ranging from about 200 million years for Olympus up to 2 or 3 billion years for the collapsed giant known as Alba Petera ("White Pan"), located northeast of Olympus.

Geologists point to the similarity of the Tharsis Dome to the Hawaiian Islands chain on Earth. Both are the result of volcanic activity, pinpointing a deep "hot spot" from which molten rock is pushing up. The resulting mountains are remarkably similar in shape, forming what are

Mauna Loa, a shield volcano in Hawaii that rises 13,680 feet above sea level, is similar in its gentle slope to the Martian shield volcanoes (author photo).

called "shield volcanoes," smooth, gently sloped mountains made from freely-flowing lava, not the viscous, explosive lava of a Vesuvius or a Mt. St. Helens. But the Hawaiian Islands are a long chain of volcanoes, extending all the way from Midway Island at one end to the Big Island of Hawaii at the other. This long chain results from the slow movement of the Earth's skin, which in the mid-Pacific is sliding northwestward. The old islands are at the tip of the chain, while the youngest, the Big Island, now sits over the hot spot and is still growing. Its highest volcanoes, Mauna Loa and Mauna Kea, rise over 13,000 feet above sea level and about 20,000 feet above the ocean floor.

Mt. Olympus, the largest shield volcano on Mars, rises more than 80,000 feet above Mars' equivalent of sea level. This view is looking straight down. The large caldera is at the center and the high basal cliffs are at the edges of the picture (NASA).

On Mars, things are different. The Tharsis Dome lies over a hot spot, but there is no movement of Mars' skin – no "plate tectonics," as geologists call it. The Dome has been in the same place for billions of years and has grown to immense dimensions. Mt. Olympus, to one side of the Dome, is 80,000 feet high above Martian sea level, nearly three times as high as Everest. It is 500 miles across at its base. Were it set down on the US, it would smother all of New England, from New York to the northern tip of Maine.

Compared with Mt. Everest, Mt. Olympus would not be a particularly imposing site from its base, in spite of its great height. In fact, a climber would spend much of the time climbing the peak without being able to see the summit, which would be beyond the horizon. The top of Everest is

The cliff at the base of Mt. Olympus is as high as many famous mountains on Earth. This view from above shows the high slopes of the mountains on the right and the flat plains below the cliffs at the left (NASA).

nearly always visible (weather permitting) from near its base, as it rises 13,000 feet above the traditional base camp on the Khumbu Glacier, only 9 horizontal miles from the summit. Olympus' summit, though it is more than 60,000 feet higher than its base camp, nevertheless lies 150 miles away, out of sight beyond the nearby rocky slopes.

The great wall

Though the summit can't be seen from the base camp for Mt. Olympus, it does have a remarkable and intimidating view. A great basal cliff surrounds this massive mountain, averaging about 10,000 feet high. This steep cliff forms a pedestal upon which Olympus sits, and is the first obstacle for any climbing expedition.

Olympus' remarkable basal cliff is a major Martian mystery. Terrestrial geologists have many guesses about what might have caused it, but no one knows for sure. Perhaps your climbing party will have time to do a little investigating while camped at its foot and will find evidence to solve the mystery. If you do so, the first clue that you will notice is that the cliff face shows layering, indicating that the various sheets of lava that flowed from the summit over time were abruptly sheered off when the cliff was formed. But how?

One possible explanation is that it resulted from an elevation of the entire mountain, which was pushed up from below, leaving its edges as a cliff formed at a circular fault zone in the Martian crust. Another idea is that erosion (most likely wind erosion, as there is no liquid water on Mars at present) wore away softer, older surface material, but left the harder volcanic rocks of the mountain. A third idea is that the lower flanks of the mountain simply slid away as landslides, spilling the rock out over the surrounding plains. A more exotic suggestion is that Mt. Olympus' pedestal was formed at a time when it was near the Martian north pole. Mars wobbles on its axis, as does the Earth, and there is evidence that the Tharsis Dome was near the pole when Mt. Olympus was being formed. At that time, the volcano would have been under the northern ice cap. A volcano that forms under a glacier will have a different history from those on dry land, especially at its edges, where the lava melts the ice. The gentle slope of the upper reaches of the mountain will change to a steep slope at the bottom, where the melting ice will break up and carry away the hot lava rocks, leaving an abrupt edge.

Maybe there's another explanation, not yet considered. The puzzle will probably only be solved by a team of adventurers, such as yours, that makes a careful study right there at the base of the cliff.

Establishing a base camp

For your first climb of Olympus, it is probably best to avoid the challenge of the steepest parts of the basal cliff. We know that it can have an average slope as steep as 35° from the top to bottom, but there probably are much steeper sections at various levels, not yet seen by Martian spacecraft. A climbing party ascending the cliff might be delayed by days if it encounters a long, unbroken vertical edge of a lava layer, hundreds or thousands of feet high. Therefore, it is recommended that the ascent be made from either the south or the east, where there are places where the cliff has a more gentle slope. A preferred base camp position in the east would be on the smooth high plains at longitude 129.5° west, latitude 21° north. From there the distance to the summit is at a minimum, only 150 miles in a straight line, but this means a somewhat steeper climb in total. However, except for the basal cliff, which this route avoids, steepness is not the main problem at Olympus, distance is. Even for this relatively short route, the entire operation will take nearly a full Earth month.

Speaking of months, before describing the route and plan, it is good idea to review the kinds of days, months and years that one finds on Mars. The days are not a problem. Mars turns on its axis in just over 24 hours, so that we Terrestrials will feel quite at home with the passage of day and night.

Months on Mars are another matter. There are two moons, Phobos and Deimos, both much smaller and nearer to their planet than our Moon. Because it is so distant, about 250,000 miles from the Earth, our Moon orbits slowly, making a complete circuit (with respect to the Sun) in 29 days, the basis for our month. Phobos is a little less than 6,000 miles from the center of Mars. It orbits the planet so fast that a Phobos month is only 9 hours long. Three times each day it zips up above the western horizon, speeds across the sky and sets in the east. Although a very small moon, only about 15 miles across, its proximity to Mars means that, as seen from the Martian surface, it will look almost as large as our Moon as seen from the Earth.

Deimos is farther away, a little more than twice as far from the center of Mars. It orbits more slowly than Phobos, with a period just over a

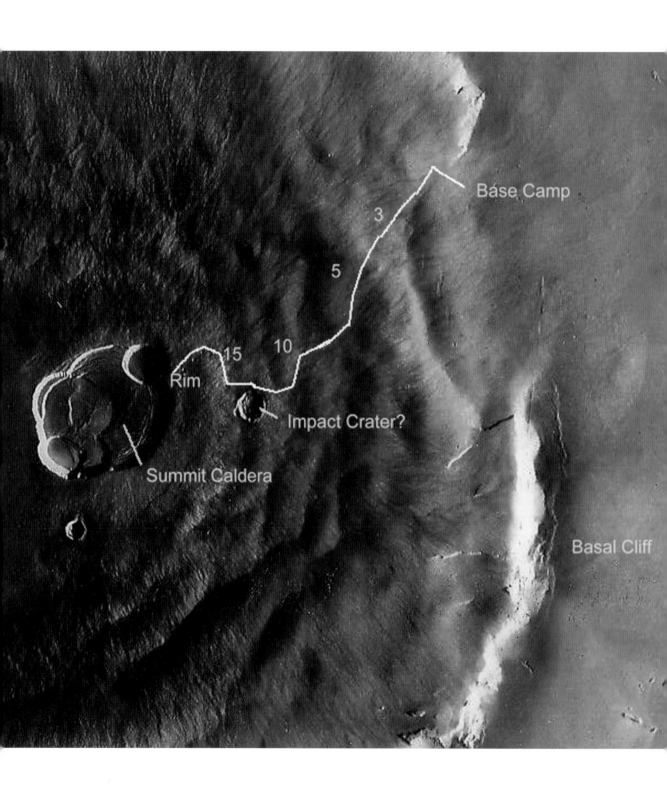

Martian day. About half Phobos' size, Deimos appears as a small, dim moon that creeps slowly across the sky from east to west.

The Martian year is nearly twice as long as Earth's, a result of its greater distance from the Sun, meaning a smaller solar gravitational pull. If any members of the climbing expedition are on salary, it might be a good idea to stick to terrestrial time. Otherwise, those paid on a yearly basis will be badly underpaid on Mars, while those on a monthly salary will get paychecks every nine hours!

Mars has seasons like Earth's because its axis is tilted by nearly the same amount as Earth's. It is best to plan a climb of Olympus for the Martian summer to take advantage of the longer and warmer days. Even in summer, however, the Martian cold is severe, with expected daytime temperatures never above 32 °F at base camp, and nightly lows near −200 °F. High on Olympus, the temperatures will be much lower.

Supplying a Mt. Olympus climbing party requires a lot of advanced planning. Of course there are no Swiss chalets along the way, no trout streams to fish, and no water to add to a mountaineer's dried macaroni and cheese. Food, water and oxygen must be brought along. And there are no yaks to carry it and no Sherpas to lead the yaks and to find the climbing route. Carrying enough food and water for a month's excursion in a backpack would be impossible even in Mars' lower gravity (two fifths of Earth's), especially for a climber wearing a cumbersome Mars Environment Suit. The solution to this problem will not be easy.

Just as it is now considered reprehensible for climbers of Everest to be carried or "short-roped" to the summit, so is it considered cheating to climb Mt. Olympus accompanied by motor vehicles carrying supplies. And it would hardly be considered a first ascent if you were to send someone on ahead in a Martian rover to leave food in caches along the route to the summit. Air drops are considered OK for some of the more lengthy terrestrial ascents, but Mars' thin atmosphere doesn't support normal aircraft. An ordinary plane or helicopter – even a U-2 – simply can't get off the ground, even at Mars' sea level, because the atmospheric density is so low.

One possibility would be to have several small rockets launched ahead of time from a Mars space base. If each carried a payload of food for five days, six of them would be enough, spaced on the slopes so that you would reach them on both the ascent and descent.

Another idea is to use a remote-controlled blimp. A huge, cumbersome one will be required to carry your materials, as even a large volume of

A lava tube in Oregon, looking out of an opening formed by collapse of part of the ceiling. Similar lava tubes have been found on the slopes of Olympus, especially near the lower northeast flanks (author photo).

hydrogen has limited carrying power on Mars. The advantage of a blimp, however, is that it can be steered close enough to the ground that your well-padded supply boxes can be dropped from only a short distance, decreasing the chances of damage.

In either case, it must be remembered that refuse must be removed from Mt. Olympus following the climb. Experience at Everest in the twentieth century vividly and disturbingly demonstrated the importance of such a rule.

The plan

Here is a possible day-by-day plan for the ascent from the eastern base camp position. Of course, the size of the group, the nature of the equipment, the physical condition of participants, and individual taste will dictate departures from this basic plan. On the ascent it is best to plan a daily distance of only 10 miles or so, allowing time for exploration, geological research and any unforeseen difficulties. For the descent 15 miles a day probably will be possible.

Day 1. Leave base camp as the Sun rises in the east. Head northwest, past several small craters (are they meteorite craters or volcanic vents?), towards a smooth breach in the basal cliff.

Day 2. After camping near the crest of the lava flow that tumbles over where the cliff should be, continue up, now to the southeast.

Day 3. Make camp near the base of a moderate slope. Explore the "mysterious" linear feature to the south — probably a lava tube or channel.

Days 4 and 5. Camp on fairly steep lava fields. Here the slope averages about 25°.

Days 6 and 7. Turn more southward along the foot of a lava rise to the west.

Day 8. Camp in a valley that leads southwesterly up through the next lava step.

Days 9, 10 and 11. Follow the wide lava valley up to the base of a shoulder. The altitude of Camp 11 is 60,000 feet!

Days 12 and 13. Pass around a lava mound to the north and then ascend to the northern rim of a large impact crater, about 9 miles in diameter. From Camp 13 at its rim, look down at its patterned floor. Steep walls will make a descent hazardous, but perhaps someone in the party will want to explore at least partway down. Morphologically this crater is somewhat different from volcanic craters and vents. It has a sharply raised rim and a bowl-shaped floor, although this example seems to have a lot of debris on its floor. Look for samples of impact-deformed rock, such as impact breccia, a coarse mix of fragments of rock and melt that is formed when a meteorite hits the ground. If you find nothing but volcanic rocks instead, we may have to revise the claim that this feature is not a volcanic side crater of Mt. Olympus.

Day 14. Camp near the base of a steepish slope that is marked by linear lava flows and several intriguing small crater-like features.

Days 15 and 16. Skirt around the northern side of the slope to the west, the last steep rise before the summit.

Day 17. Pass upwards along a row of small pit craters, about a mile across. These are probably similar to the ones found in Hawaii Volcanoes National Park along the Chain of Craters Road, but the Martian ones are much larger.

Days 18 to 20. Reach the summit rim. Stretched out below you is the giant summit caldera, many times larger than terrestrial examples such as that on Mauna Loa, though very similar in structure. The caldera is a complex of frozen lava lakes from which molten rock issued during the periods of the building of the mountain. At least six different lakes, frozen at different levels, can be seen, but the vastness of the caldera, nearly 50 miles across, permits only a part of it to be appreciated from the rim. The walls are very steep and the floor is marked by channels, wrinkles and cliffs between lake levels. To the north is one of the two deepest and most recent lava lakes, about 9 miles across. Off to the southeast is the other deep lake, which is less round. Three days are scheduled here to allow time for exploration along the rim and possibly a little ways down into the caldera. The best route down is probably a comparatively gentle slope just to the west of the edge of the northern, deep lava lake.

Days 21 to 32. Descend to base camp, following the route up and revisiting the supply caches for the food that was left in them. At base camp

The complex caldera at the top of Mt. Olympus (NASA).

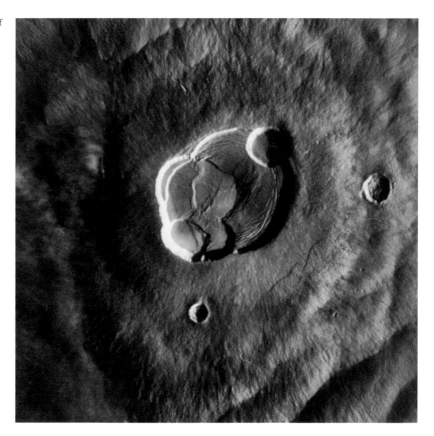

celebrate having climbed the highest mountain in the Solar System. But be careful when you open the bottle of champagne. At Mars' low atmospheric pressure, the cork can become a dangerous missile!

2

Higher yet? Other high peaks of Mars

A recent news story claimed that Olympus is not the highest mountain on Mars. While there is no argument that Mount Olympus is the biggest volcano known to science, this claim suggested that one of three other volcanoes that lie more centrally on the Tharsis Dome might have a slightly higher summit. However, from a scientist's point of view, the important measurement is the height of a mountain above its base, not above sea level or a similar reference level. Mauna Loa, for example, is only 13,677 feet above sea level, but nearly 20,000 feet above its base. A scientist considers it to be a higher mountain than Mt. Rainier, though a mountain climber considers Mt. Rainier (14,410 feet above sea level) to be higher, even though it rises only 12,000 feet above its base. Similarly, even if it were true that one of the other Martian volcanoes reaches an elevation above "sea level" that is higher than the top of Mt. Olympus, its height above its base is much smaller.

We don't hear very much about the other Martian volcanoes. The reason probably is partly the fact that Olympus is much more massive, is

The terrestrial volcano Mt. Rainier rises 14,410 feet above sea level. The contrast between this scene and the views of Mars' volcanoes emphasizes the great environmental differences between the two planets (author photo).

The Martian volcano Ascraeus rises about 85,000 feet above the Mars equivalent of sea level. This imaginary view shows it surrounded by the Tharsis plains. Mars' pink sky is colored by red airborne dust.

higher above its base, and occupies a much larger area. It also has a more familiar name. It was named a century ago by astronomers who saw a white spot at its position, usually in the winter, leading them to infer, correctly, that a high mountain exists there. They named the feature Nix Olympica, the "Snows of Olympus," after the Greek mountain that was the home of the classical Greek gods. We now know that the white spot was probably clouds at the summit, not snow.

There are three other high Tharsis volcanoes, discovered when Mariner 9 arrived at Mars to map its surface in 1971. Arriving during a global dust storm that obscured most of the Martian surface, the spacecraft sent back alarmingly blank pictures, except for the Tharsis area, where the tops of four high mountains peeked out above the haze. The easternmost of these was named Ascraeus Mons (Mt. Ascraeus), after the name given that portion of Mars by nineteenth-century astronomers. The other two, Mt. Arsia and Mt. Pavonis, were also named after traditional features on the old maps.

The peak at the top of the dome

Mt. Ascraeus lies near the highest portion of the giant Tharsis Dome. Its base is at an elevation of nearly 30,000 feet, about 18,000 feet higher than the base of Olympus, which is perched near the western edge of the dome. The bottom of Ascraeus is at nearly the same elevation as the summit of Mt. Everest. Its top is almost as high as that of Mt. Olympus, perhaps even a little higher, about 85,000 feet above "sea" level.

Ascraeus is an easier climb than Olympus. You start at a higher elevation and the distance to the summit is smaller. But no expedition on Mars is easy. Climbing Mt. Ascraeus will require almost as much advance preparation, effort and time as climbing its bulky neighbor to the west. Geologically it is pretty similar. It's a shield volcano, built from thousands of eruptions of runny, low-viscosity lava that issued from its central crater.

Where should you start? Choosing a route up Ascraeus presents a problem, but not because of difficulties like those presented by Olympus' basal cliff. Instead it's a matter of deciding among three different but interesting approaches. If base camp is established on the northwest flank of the mountain the party can explore the remarkable aureole that fans out from the bottom of the mountain. Marked by arc-shaped ridges, the aureole is something of a mystery. One interpretation is that it is a glacial moraine, material left long ago, when thin glaciers may have

Glacial moraines on the Earth being formed below the Gornergrat in Switzerland. The centers and edges of these active glaciers are covered with rocks and dirt brought down by the glaciers' carving action. This morainal material is left at the bottom of the glaciers, forming a terminal moraine. The arc-shaped ridges below Ascraeus may be examples of terminal moraines (author photo).

A U-shaped valley on Earth, carved by a glacier that has now receded. No such evidence for glacier-carving has been seen on the slopes of Ascraeus (author photo, Yosemite Valley).

enveloped the mountain. As the hypothetical ice sheet shrank, it may have left the concentric ridges of the aureole, which strikingly resemble those seen in terrestrial moraines of retreating glaciers. This may seem preposterous at first, considering that there is no sign of present or past glacial activity on the heights of the volcano. However, as mentioned in Chapter 1, it is quite possible that the Tharsis area was long ago near Mars' north pole, at which time ice would be quite likely. Perhaps there was enough water vapor for ice formation only in the thicker air near the base, giving Ascraeus and its companions ice skirts. Or possibly the glaciers went higher at one time, only to have the tell-tale evidence of their existence (U-shaped valleys, cirques and lateral (side) moraines) covered over by later lava flows. A northwest approach to the mountain could give your expedition a chance to look for geological clues to solve this aureole mystery.

An approach from the south could also present some opportunities for interesting exploration. A huge triangular valley cuts into the mountain's southern lower flanks and from it spreads an immense lava fan, extending as much as 120 miles from the mountain. The top of the valley is some 20,000 feet higher than its base and its walls are marked by huge gullies. The surface of the large, flatish lava fan has numerous craterlets and channel valleys. An adventurous team might choose to climb the mountain by first exploring these and then ascending straight up the valley, checking to see exactly how it was formed.

The route to the summit

A slightly easier but no less interesting ascent can be made from the northeast. On this side of the mountain the lower flanks are rife with lava channels, collapsed lava tubes, and other features, enough to be a volcanologist's field of dreams. The extra interest these can provide on the expedition may lead you to choose this route. A suggested itinerary, based on the assumption of a fit and well-equipped group, follows. Among your navigational instruments, be sure to take one that will allow you to measure accurately the altitude of the highest point on Ascraeus so that the dispute about which mountain is the Solar System's highest can be resolved.

Base Camp. A good location for the base camp is the flat plain at the northeast foot of the mountain at latitude 13 °45' north, longitude 102 ° west. A variety of interesting features surround the area. There are straight rilles (valleys) that run mostly north–south, probably lava channels or partially collapsed tubes that follow the north–south trend of faults in this area of Tharsis. There are also long, narrow sinuous lava channels, some of which are connected to the straight valleys, and long lava flows, some with lava levees (higher edges, like those we build on Earth to prevent rivers from flooding). Just above base camp these features coalesce into an intricate maze of deep and narrow valleys, forming the most difficult but geologically interesting part of the trip.

Day 1. Getting an early start, pick your way carefully between the myriad lava channels to the southwest of base camp. The first stop is in the midst of these complex volcanic features.

Days 2 and 3. Continue up, staying to the south of the more complex channel fields. Camp 3 is on a nearly level shelf at 60,000 foot elevation.

Days 4 and 5. Turn to the northwest to follow the shallow shelf, passing through at least one large channel but avoiding a cliff or steep slope to the southwest.

Days 6 and 7. Skirting around the cliff, head south, ascending gradually. Camp 7 is at 75,000 feet.

Days 8 and 9. Head due west from Camp 7 to the base of the final steep slope. Day 9 is a tough one; ascend straight up the summit slope to its top, where the mountain levels out to a more gentle approach to the rim.

Day 10. Proceed to the summit at the edge of Ascraeus' huge, complex caldera. Current maps do not show unambiguously where on the rim the highest point lies, so your climbing party may need to scout around a bit to find the appropriate place to set your flag and take the traditional summit photos.

Days 11 and 12. Explore the caldera area. As at Olympus, the walls of the caldera are steep and treacherous. The main caldera is huge: 25 miles across and 12,000 feet deep. Looking down from the rim, you will see a crater almost three times deeper than the Grand Canyon in Arizona. The walls are very steep, especially on the southeast side, where a series of narrow landslides has left ridges that extend almost all the way to the bottom. Elsewhere around the edges there are huge arc-shaped faults, marking where the land slumped down when the lava lakes withdrew. The bottom of the main caldera is quite smooth.

As at Olympus, the caldera at Ascraeus is complex. The main crater is extended to the northeast and southwest, where floors of older lava lakes are partly preserved, and there are two conspicuous smaller side-craters. These are older than the main lava lake, which bisects them, leaving their smooth partial floors cut off thousands of feet above the most recent lake.

How recent was this lake molten? Is Ascraeus Mons extinct like Mauna Kea or could it still be active like Mauna Loa? Geologists don't agree on the question of whether Mars is still geologically active, but the evidence for Ascraeus Mons indicates that it has been quite a long time since its most recent mountain-building eruptions. By counting meteorite impact craters and by knowing the approximate rate of occurrence of meteorite impacts onto

[opposite] This imaginary view shows Ascraeus in eruption and surrounded by the lava flows that formed the Tharsis plains. These events probably did not all happen at once.

Mars, it is possible to estimate that Ascraeus Mons is about a billion years old. Estimates of the time since the last liquid congealed in the main caldera are about a few hundred million years. Thus it is vastly older than Mauna Loa, but is about the same age as Olympus Mons.

Days 13 to 20. Descend to base camp.

This plan assumes a very athletic and experienced team of explorers. The ascent could well take twice as long as proposed

A flow of ropy pahoehoe lava in Hawaii. This may be similar to what is found on the slopes of Ascraeus (author photo).

here, especially if it turns out that the slopes of the mountain are rougher than we can see on available images. If the route ascends through jumbled, rough lava flows, as may well be the case, progress could be painfully slow. A field trip to lava flows on Mauna Loa or its sister volcanoes would be a good idea before leaving for Mars.

The other high peaks of Tharsis

Olympus, to the side of the Tharsis Dome, and Ascraeus, at its top, are the two youngest massive volcanoes on Mars. The other two high Tharsis volcanoes are older, but both are about the same height above reference level as Olympus and Ascraeus. Further measurements are needed to find out which of the four is truly the highest. Mt. Arsia is about 2 billion years old and Mt. Pavonis just a little younger. The lava lakes in each were molten more recently, perhaps in the last few hundred million years.

These ages are fairly uncertain. For instance, geologists' different models for the cratering rate give a range for Arsia's central shield of 600 million to 3 billion years. Field-based exploration on the mountains themselves could help to resolve these uncertainties.

Arsia, Pavonis and Ascraeus, rising above the top of the giant Tharsis Dome, lie along a line running from the southwest to the northeast, each peak separated by about 500 miles. Arsia is remarkably different from the others. It has a huge caldera, 80 miles across, taking up a good fraction of the mountain. Two deep lava fans interrupt the otherwise smooth slopes, one at the northeast and one at the southwest. Lava flowing from the vents at the top of these fans can be seen to have spread out across the plains below, extending more than 200 miles from the base of the peak. A huge aureole extends equally far across the surface to the northwest. As in the case of Ascraeus, the nature of this aureole is quite mysterious.

Mt. Pavonis is rather similar in shape to Ascraeus. An unusual feature at its top might tempt explorers to ascend it. In addition to the deep, steep-walled caldera, there is an off-center circular summit plane to the northeast. Like a dinner plate, this 60-mile diameter feature has an upturned edge. Apparently it was once the floor of a summit caldera and its walled edge was once the rim of the mountain. Subsequently a higher peak was formed to its side and the latest caldera cuts across its southwest rim.

There are many more volcanic peaks on Mars, both in the Tharsis region and in other regions of Mars. Thousands of volcanoes have been identified, most of which are small, more on the scale of terrestrial examples. A few, such as Mt. Elysium, far to the west of the Tharsis region, are high and rugged and also would be well worth an expedition.

3

Descent into the Martian deep

The mysteries of Mars have long fascinated us Earthlings. Glimpsed only sketchily through our telescopes, its surface detail has had a tantalizing effect on our imaginations. Before spacecraft reached it and mapped it, the surface of Mars was a great Unknown Land, which was endowed with some wondrous features and outlandish creatures. Among the most intriguing and the apparently most well-established of these features were the famous canals.

The canals of Mars

More than a hundred years ago the world was electrified by news from Arizona about the remarkably intricate system of canals seen on Mars. First glimpsed several years earlier by the Italian astronomer Giovanni Schiaparelli and called by him "canali," these thin, dark straight lines were now being mapped in great detail by the American Percival Lowell, who had built a large private observatory on a rise that he called "Mars Hill," near Flagstaff, Arizona. His purpose was to observe the planets. Much of his personal fortune was spent on establishing this high-altitude observatory, where the unusually clear and stable air would give him the best possible views of planetary detail.

Together with his assistants, who included the astute A. E. Douglass and the remarkably difficult Thomas Jefferson Jackson See, Lowell devoted the last few years of the nineteenth century and the first few of the twentieth to the pursuit of Mars. The result was an amazing discovery: Mars must be inhabited with intelligent beings! Not only were they intelligent, he concluded, but they were doomed.

In his wonderfully persuasive book, *Mars as the Abode of Life*, Lowell explained how he deduced these facts. First, his maps of the canals showed long straight lines that were often double (with a median like a modern freeway) and that met each other at small, round dark areas that he called "oases." To him it was obvious that the canals were artificial — they hardly looked like any natural phenomenon — and he concluded that, considering their global extent, they must have been built by highly skilled engineers. Furthermore, in view of the fact that Mars' atmosphere

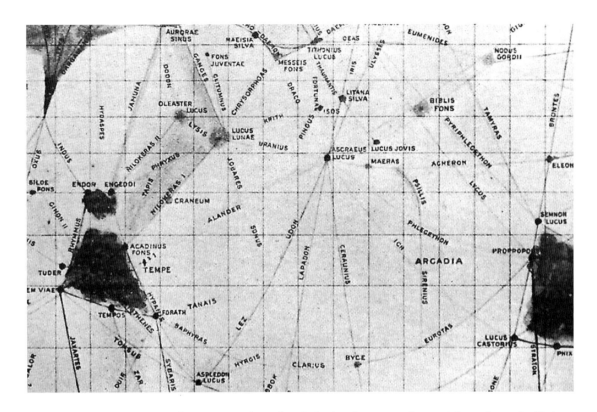

Percival Lowell's map of some of the canals he saw on the Martian surface (from the *Lowell Observatory Bulletin*).

almost never has clouds in it, these canals must have been built in a desperate attempt to save Martian civilization from extinction, bringing water from the polar ice caps to the parched and sun-baked planet. Lowell assumed that Mars was an ancient land, drying up and choked in dust and that the doomed Martians were making a valiant attempt to carry on.

It was a romantic idea, which caught the whole world's attention. It spawned uncountable stories and novels and inspired hope and intrigue and, in some spectacular cases, fear. But it also inspired skepticism. Many, perhaps most, professional astronomers remained profoundly doubtful about Lowell's canals. Their reason was simple: when they looked at Mars they didn't see them. Gradually during the first half of the twentieth century, Mars fever cooled down and the skeptics seemed to prevail. Physiologists and psychologists made experiments that suggested that the human brain, when faced with attempts to perceive very fine detail, sometimes would see lines, like a "connect the dots" puzzle. This is the brain's rather sophisticated way of doing what your computer program does when you use software to "sharpen" an image.

Mars without canals – the view returned from Mars-orbiting spacecraft Mariner 9 (NASA).

Lowell died in 1916 but his canals did not die definitively until 1971, when Mariner 9 was placed in orbit around Mars. It photographed the surface, showing mountains, craters and valleys, but not showing any long, narrow, straight features like Lowell's canals.

Except for one. Along the equator, east of the giant Tharsis volcanoes, Mariner 9 discovered a huge, nearly straight canyon system that extends for 3,000 miles. It is most definitely not a canal. No water flows in it, nor is liquid water found anywhere on Mars. Lowell was right about it being a desert-covered planet, but it is even drier than he thought. Even the meandering stream beds that Mariner 9 discovered, probably once formed by flowing water, are now dry. With the canals now recognized as physiological artifacts, Lowell's fancied Martians no longer had to be responsible for them. Martians may once have lived, but only when Mars had a moister climate and only as microscopic organisms (the possible discovery of fossils of such Martian organisms was reported in 1996).

Unlike the Martian canyons, the Grand Canyon of Earth was formed by a river as it eroded a slowly rising land mass (author photo).

But what about Mariner 9's great canyon? As Mars' only giant straight line, is it one that Lowell actually saw? The answer seems to be no; of all the hundreds of features mapped by Lowell, Mariner 9's canyon was not included. Almost all of the early telescopic features of the surface turn out to be merely differences in the darkness of the land. The actual topographic features, the mountains, the craters and the valleys, are invisible except from up close.

A grander Grand Canyon

A view from Mars' orbit of its great canyon system, called Coprates Canyon, not only shows it but reveals its astonishing proportions. The main section, named "Mariner Valley" after its discovery spacecraft, is four times as deep as Arizona's Grand Canyon. From the south rim to the north it is five times as wide, and from one end to the other it is ten times as long. And, of course, it is far more remote.

This chapter introduces a plan for exploring this remarkable canyon and for climbing its huge, steep walls. Unlike its Arizona competitor, it has no trails to follow and no mules to take you down and bring you back up. Furthermore, there is no water, hardly any air, and not even any good maps. But the challenge is exciting. The scale of things is stupendous and the journey offers the opportunity for remarkable discoveries. While we

The Mariner Valley on Mars was formed by giant faults and subsidence. This full view of Mars shows how it lies conspicuously to the east of the Tharsis Dome, the volcanoes of which are visible near the left edge of this picture (NASA).

know that Arizona's Grand Canyon was formed by the erosive action of the Colorado River, for the Martian canyon the origin is shrouded in mystery and doubt. Most astronomers agree that the complete solution will only come when an expedition such as this one brings us face-to-face with the canyon walls.

Some serious faults

The simplest explanation of the canyon system is that giant cracks opened up during the formation of the nearby Tharsis Dome. These cracks, called faults by geologists, might be analogous to those that formed the Great Rift of East Africa, which is about as wide (but nowhere nearly as deep). The grand scale of the Martian canyons suggests that parallel faults may have broken the crust of the planet when submartian pressures started to build up the Tharsis area. Then sections between nearby parallel faults subsided, forming what is called "tectonic grabens."

Other ideas have also been suggested to explain the canyons. Perhaps when water was on Mars in liquid form, water erosion was involved in some way (though no one believes that it is a mere river valley, with water coursing down from the heights of Tharsis to spill out to the east at the far end of the canyon). Perhaps the strong winds sometimes detected on Mars have eroded the area, gradually wearing away the land over billions of years. Possibly all of these things happened and to various degrees they shaped what we see. Or perhaps something else occurred here on Mars. A well-planned expedition should help settle the question.

The North Rim

An expedition to a canyon's depths is best preceded by a visit to the rim. From Mars base, perhaps located in the Luna Planum ("Plain of the Moon") to the north, the distance would be about 600 miles. You should plan to drive a high-wheeled, balloon-tired Mars expedition vehicle, outfitted with enough food and oxygen to last about two weeks.

Leaving Mars base in the morning, the expedition should arrive early the following day at the North Rim. The route is across a largely feature-less landscape, but the going will be slow. There are no roads and the way is littered with large and small dust-covered rocks, most of them volcanic in origin, having been broken up and scattered about by billions of years of meteorite impact cratering.

Nearing the canyon complex, the route crosses the equator and swings to the east of the feature called the Ganges Chain of Craters, a broken series of circular depressions that extends about 120 miles, nearly east–west. The craters are probably "sink holes" associated with a crack in the surface. Or possibly they are volcanic craters analogous to the "Chain of Craters" at Kilauea on the Island of Hawaii. The Ganges Craters are much larger, however, ranging up to about 5 miles across, while the Hawaiian chain craters are only half a mile or less in diameter.

The Mariner canyon system has many of these crater chains, some much larger than the Ganges Chain and all nearly parallel to the main canyons. This latter fact is what makes them look suspiciously like smaller examples of the canyons themselves, possibly similarly formed

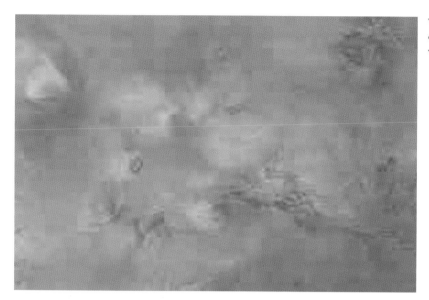

White clouds and fog sometimes cover Mars' mountains and fill its valleys (NASA).

when the uplifting of the Tharsis Dome cracked and stretched the Martian surface layers. You probably won't stop to examine the Ganges Craters, as the puzzle of their formation is probably the same puzzle that the expedition to the giant Coprates Canyon will explore.

The expedition will arrive at the North Rim of Coprates at dawn. The slightly anemic sun, about half as bright as seen from the Earth, rises in a pinkish Martian sky. As the sunlight penetrates deep into the canyon, you may see a light misty fog in the depths, which quickly evaporates in the warming sunlight. This is probably water vapor, formed in the same way that fogs on Earth form in chilly valleys, to be dispersed by the morning warmth. Soon the expedition will know its nature better, because in a few days the exploring party will be down in those foggy depths.

As the view clears, the canyon's grand scale and magnificent proportions will become almost overwhelming. The bottom is more than 20,000 feet below the rim here at Coprates, four times farther below you than the floor of any terrestrial canyon seen from its rim. Powerful binoculars can bring it closer and you can scan the area beneath you where the base camp will be established. It looks inviting, smooth-floored and solid. There is no river meandering through it, there are no trees and no roads, but then neither are there any rattlesnakes or grizzlies.

Across the canyon, about 75 miles away, your binoculars will bring into focus a sharp-edged ridge that forms a sort of backbone along the middle

As on Mars, ground fog sometimes forms in the bottom of valleys on Earth (author photo).

of the Coprates Canyon. This ridge divides the Canyon in two, but off to the east, towards the rising Sun, it gradually melts into the true South Rim. There, remnants of the smooth surface of the plains top this ridge, but in front of you the original surface is gone, having eroded away down into the depths. Beyond the ridge-top, just visible from your vantage point, you may be able to make out the South Rim, about 100 miles away. But it may be below the horizon and invisible from where you'll be standing. The curvature of the surface of Mars is greater than for the Earth (Mars is about half as large), so that far-away objects sink below the horizon at closer distances than on Earth.

Looking to the west, however, you will be able to make out the true South Rim. The central dividing ridge gradually dissolves down into the floor of the canyon in that direction. At a point about 130 miles west of your vantage point it is gone completely and the South Rim may be just barely visible in the haze. There the canyon is at its widest, 170 miles from rim to rim. That portion of the Mariner Valley system is called Melas Canyon and soon the expedition will be picking its way along the floor on its way to base camp.

Down the Giant Slide

After a day of viewing and exploring at the North Rim and a quiet night in the camper section of your vehicle, it is time to drive down into the canyon. This is not a simple thing to do. The steep walls of Coprates provide no easy way down. Instead, you will have to backtrack almost to the Ganges Crater Chain, where in this land of giant things there is a giant landslide. The northern edge of Ophir Canyon, a side canyon of the Mariner system, spills down into the depths in one of the largest land-slides known in the Solar System. The top of the slide is about 30 miles across and its full length is 120 miles. Descending smoothly into the Ophir Canyon, it doesn't reach the true bottom until it spills half-way across the Candor Canyon, an adjacent, blunt, deep valley.

You can probably allow 15 hours of driving to negotiate the fairly easy trip to the top of the landslide, arriving after sunset on Day 3. Next morning, as soon as sunrise allows, you should take a close-up look at the Giant Slide. From orbital images returned to Earth it looks like a snap: a nice, smooth, gentle route to the bottom. No terrestrial rock slide is much fun to drive on. The rock and debris that slid and fell to form the slide are

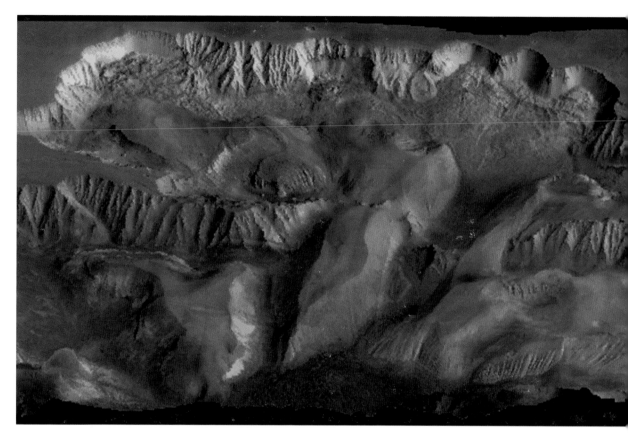

usually unstable and even a small new disturbance can cause more sliding, with devastating results.

However, the Giant Slide has been here for something like a billion years or more. Marsquakes and Martian weather have probably jiggled away most unstable sections of the landslide and your way will probably be clear and your trip down uneventful. You should allow a good 12 hours to negotiate the 120 miles to the floor of Candor Canyon.

South Candor Canyon is a complex part of Mars' great rift valley system (NASA).

The floor of Candor

At sunrise on Day 5 you can look around at last at the floor of the canyon to see what clues there are about its origin. When the mists, if any, have cleared, your view will be spectacular. Rising steeply to the east is the sheer wall of the canyon. Behind you to the north is the long, smooth, gentle slope of the Giant Slide. To the west is a complex of ridges that sepa-

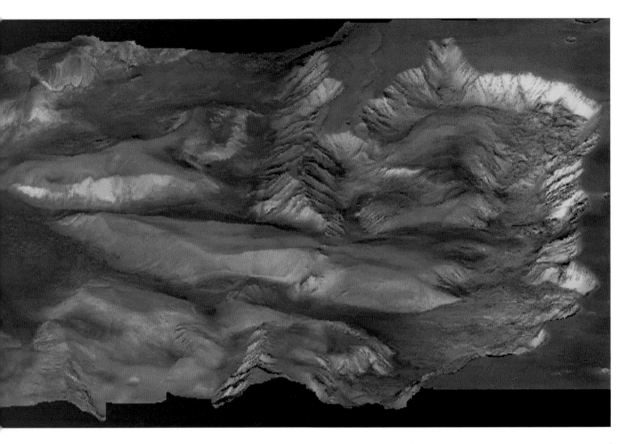

West Candor Canyon as seen in an oblique view. The expedition will spend some time in this canyon exploring it for clues about the formation of the complex Martian canyon system (NASA).

rates Candor from Melas Canyon, and to the south is the wide opening of the canyon through which you'll soon proceed.

Climbing out of the truck onto the surface in your environmental suits, you will want to explore Candor Canyon's floor briefly before breakfast. You'll have more time at Coprates base camp to study the canyon floor problem, but at least a quick survey of Candor is called for, as it is one of the best places to see a mysterious series of rocks called the "layered deposits." The layers show up as mid-canyon hills, covered with dust. They are also seen well in parts of Hebes and Ophir Canyons to the north. Their origin is unknown and your photographs and the samples you collect may help to narrow the possibilities. Some geologists suggest that they are lake-bottom deposits. Perhaps long ago there was enough liquid water on Mars to fill the bottoms of the canyons with lakes. The observed horizontal beds of rocks would then be the hardened silt layers formed at various stages of the lakes' ancient history. The water may have

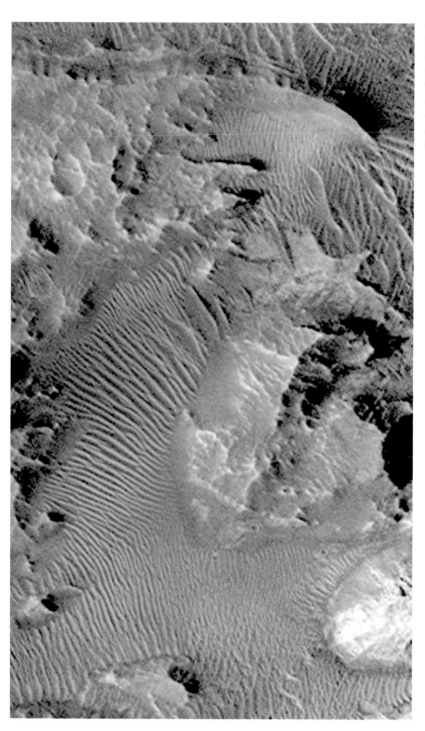

This high-resolution image of Hebes Canyon, showing the sand dunes in its floor, was obtained with the Mars Global Surveyor. This is probably the kind of countryside that will be encountered frequently as the expedition moves among the canyons of Mars (NASA).

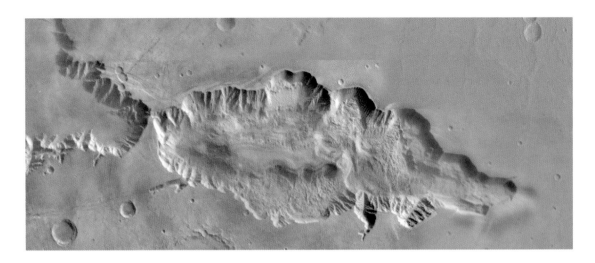

Hebes Canyon, located to the north of the main canyon system, provides a well-studied example of a sand-covered canyon floor (NASA).

seeped out from the water-rich layers of rock that we know were (and possibly are) found beneath the Martian surface – presently as a sort of permafrost. This water was melted either by volcanic heat or a warmer Martian climate and, as the canyons formed, it may have cascaded down from the steep walls, forming giant, shallow lakes. Possibly these canyon lakes were inviting enough for primitive life forms to develop. You will surely collect enough specimens of rock to take back for analysis to check out this possibility.

But there are other possible explanations of the layered deposits. They may simply be different layers of volcanic rock, laid down as the Martian surface was gradually paved by molten lava. Or they may be layers formed by the erosional action of the winds of Mars. We have witnessed from Earth the immense wind storms that can circle the Martian globe at times. The dust carried by the winds can help erode the Martian surface, scouring out holes and channels and then depositing itself in layers. Or there may be other causes. Your quick look and hastily collected rock samples may provide some vital clues.

But you cannot tarry at Candor Canyon. The great adventure of this expedition lies ahead. It is both a scientific adventure and a climbing adventure. You will attempt to scale the forbidding cliffs of Coprates, ascending through the history of Mars step by step.

4

The cliffs of Coprates

As we have seen, the vast Mariner Valley system is made up of many canyons. The straightest, steepest and deepest is the Coprates Canyon, the final destination in the second part of your third Martian adventure.

East to Coprates

From the foot of the Giant Slide, your route to Coprates takes you through a wide gate and then across a vast, flat, deep canyon floor. Between Candor Canyon and the broad canyon to the south there is a breach in the walls through which you must guide your vehicle. The distance to the base camp is 120 miles and all day will be needed, considering the many stops that you'll want to make to explore what you find along the way.

One of the first stops will be at the brink of a remarkable small valley. From orbit this looks almost like a meandering stream bed. It seems to issue from the base of a steep cliff on the north side of Melas Canyon and then winds its way south for about 100 miles, finally disappearing at a gentle, low hill near the canyon's center. Is it a dried-up river bed like the famous Nirgal Valley 1,200 miles southeast of here? There are many such meandering valleys on Mars and their nature is still controversial. Most scientists are convinced that they were formed by water at a time when Mars was warmer and had a thicker atmosphere. If water were available now, it would not remain liquid long enough to form a valley. If warm, it would evaporate; if cold, it would freeze.

As you pause at this canyon bottom, you should try to figure out its cause. Does it have water-carved banks? Are sand and rocks sifted and separated by size as in terrestrial river beds? If they are, then this feature is probably a valley formed by seepage of ground-water from the main canyon walls. Most such valleys on Mars, such as Nirgal Valley, seem to have formed by seepage rather than by rainfall. In a rainy climate the river valleys have many branches, and the branches have branches, with tributaries spread out over a large area. Even in a desert, rain-caused river valleys follow this pattern. On Mars, however, there is little branching and the tributaries are short, ending in bowl-shaped alcoves. Thus, most scientists believe that the long valleys like Nirgal, which is 500 miles long,

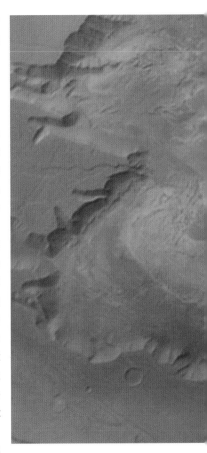

Melas Canyon on Mars is part of the complex of canyons through which you must travel to reach the base camp (NASA).

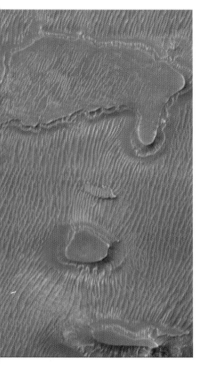

Sand in the bottom of Melas
Canyon was probably deposited
by winds (NASA).

were most likely formed when ground-water was released or ground-ice
was melted. Perhaps the much smaller stream-bed in Melas Canyon was
also formed in this way.

Continuing on, the way is smooth for many miles, allowing a good
chance to examine the undisturbed canyon floor. An important thing to
look for is evidence that this surface might be a section of the old, original
Martian surface. If the canyons were formed when faults opened up and
slices of Mars sank down thousands of feet, remains of that old surface
may still be recognizable. Elsewhere in Coprates Canyon some large
impact craters are seen, strongly suggesting that there are remnants here
of the old cratered terrain. Your close-up views will be an exciting chance
to check this idea by looking for smaller old, eroded impact craters.

By evening you will have reached the base camp position at last. Here,
at a longitude of 65 ° west and a latitude of 12 ° south, you sit at the foot of
one of the steepest cliffs on Mars. Twenty-thousand feet above you, invisible beyond the nearer ramparts, is the top of the North Rim, where you
stopped three days ago to peer down. The top is higher above you than
most terrestrial mountains rise above their bases and the task of
ascending this monstrous cliff will appear daunting.

The climb

The morning sun reaches the canyon floor long after it lights up the
heights above. Off to the south, you will see the sunlight on the serrated
mid-channel ridge, while your climbing route to the north is still in deep
shadow. As the shadows retreat, the route-finders can make more definite
plans than were possible from the Mars orbiter images. The top of the cliff
is about 15 miles away but the route up will be much longer, as there will
be rough and irregular slopes to negotiate. So that you can make your way
safely upward and do some geology along the way, you will want to plan a
total of seven days on the ascent. Like a traditional Mt. Everest expedition,
the plan will require many trips up and down to the various camps, which
must be supplied with stores of food, water and air.

Do you really need to carry water? On Everest, climbers can simply
start up the stoves in their tents and melt snow, but there is no snow on
Mars. High on the cliffs, however, there may be ice frozen into the porous
rocks. If so, your supply problem may be simplified — you can try to
extract this water.

The Mars water remains a mystery, one that you can explore not only for something to drink, but also for gaining a better understanding of Mars history. Elsewhere on Mars the geological evidence for water is common. Not only are there dry river valleys like Nirgal Valley and, possibly, the small channel that you crossed on the floor of Melas Canyon, but there are giant flood channels, evidence that in the distant past there was a period of widespread melting and catastrophic flooding. At the sources of this (presumed) water are shallow, scalloped cliffs, and it looks very much as if warming of the surface must have melted great quantities of sub-surface ice, leading to a sudden flow of water downslope and out onto the lower plains. The location and extent of this sub-surface ice, if it exists, may be found by your expedition higher up on the cliffs of Coprates.

An image that shows what you might see if standing at the bottom of Coprates Canyon below the towering cliffs and the pink Martian sky.

The talus first

When this trip was being planned back on Earth there were few good high-resolution images of the canyon walls. The surface of a steep cliff as seen from above, from orbit, is foreshortened and details are obscured. However, one can make some very plausible guesses about the nature of the foot of the cliff.

It is not expected to be like Yosemite Canyon, where the cliffs are nearly vertical all the way down to the valley. Yosemite was formed in hard, metamorphic rock by the movement of a giant glacier, which cut through the mountains like a knife, leaving smooth, precipitous walls. Streams jet out into space at the top, forming its famous high waterfalls. At the foot of the Yosemite cliffs there are small steep skirts, called "talus slopes," formed by recent erosion and made up of rocks and boulders brought down from the heights by rain and snow. The Coprates walls are different; they are not glacier-carved, and erosion there has not been caused by the continuous precipitation of a moist climate.

One also does not expect Coprates to be like the bottom of Arizona's Grand Canyon, which was carved into ancient metamorphic rocks, the famous Vishnu schist, by the grinding action of the Colorado River. While the glacial valley at Yosemite is "U-shaped," the bottom of the Grand Canyon is shaped like a "V," with steep walls slanting right down to a line where the river now flows. Most rocks and dirt that fall down to the bottom are swiftly carried away by the river, so there is no extensive talus slope. When one lands one's river raft on a sand bar, it is only a few steps to the foot of the cliff itself.

Instead, the Martian cliffs are expected to be more like those found on Earth in riverless, dry deserts or in equally water-free arctic mountains. Northern Africa, the Chilean Andes, Death Valley in California, or Antarctica are the places to look for cliffs that might be useful analogies. At the bottom of these valleys there are usually quite extensive talus slopes, broad skirts of rocks and boulders that consist of a chaotic mix of pieces of the walls above. The angle of these talus slopes is relatively gentle. With no rivers to wash them away or glaciers to sweep them up, the rocks lie at what is called the "angle of repose." On Earth, this is usually less than about $30°$, depending on the sizes of the rocks. It is the steepest slope that will form without the material slipping and sliding on down farther. On Mars the surface gravity is lower, so this angle can be

The bottom of the Grand Canyon is narrow and V-shaped, unlike the flat-bottomed Mariner Valley (author photo).

steeper than for similar talus on Earth, but you won't know before you get there exactly how steep because the sizes of the boulders, rocks and sand will not be known in advance.

The first day, therefore, will involve scrambling up through loose, possibly unstable, slopes of mixed rocks and sand, covered by reddish dust. You will need to pick a way that is not too steep to avoid setting off dangerous avalanches and this may make the route tricky to find and circuitous. The day will probably be taken up entirely by this treacherous trip and Camp 1 will be established in some nice flat area near the top of the talus, possibly about 5 miles from the base camp. Most of the climbing party will return down the next day, having served primarily as porters, carrying supplies for this camp as well as higher camps.

Water, water anywhere, perhaps a drop to drink?

Day 2 is possibly spent on more stable rock, though probably not the solid, smooth rock that Yosemite climbers enjoy. Instead it is probably ancient shattered rock, originally volcanic or igneous, but greatly altered in early Mars history by the devastating bombardment of giant meteorites. About three and a half billion years ago, the planets suffered a severe pelting by debris left over from planetary formation. This material ranged in size up to tens or even hundreds of miles in diameter and it punctured the young planetary surfaces, pulverizing the recently solidified surface rocks. The Moon provides a dramatic history of this period of massive bombardment. The lunar maria (the lava-filled "seas") were formed by the largest missiles and the Apollo astronauts brought back rocks that provide the actual dates of these events. The giant lunar impact basins were punched out and then filled by lava between 3.1 and 3.9 billion years ago.

Mars seems to have had a similar experience. In the most ancient terrain of Mars, found in its southern hemisphere, there are giant circular basins that look very much like the lunar maria. The largest example is the Hellas Basin, nearly 1,000 miles in diameter. It was probably formed during the period of major bombardment by a chunk of primitive rock about 100 miles across. The whole of Mars' surface probably was pelted by such objects, with the result that Mars' ancient surface was shattered.

On Earth, there is so much geological activity that we have little chance of finding old rocks from this period. Our surface is continually being destroyed and regenerated. But we do know what happens when a large meteorite hits the Earth. Some 200 impact structures are found scattered about the Earth's surface and many of these have been studied in depth. The impact causes a large circular depression (a crater) to form, with surface material thrown out to the surroundings by the explosive collision. The upper layers of the sub-crater rock and most of the ejected material is in the form of a mixture, called "breccia," made up of various-

The Hellas Basin on Mars is shown here as the deep blue circular feature in the lower part of this map, which is based on elevations measured by the Mars Global Surveyor. It is one of the Solar System's largest known impact basins (NASA).

Impact breccia, a rock formed by the mixing together of debris and melted rock during the impact onto the ground of a meteorite (University of Washington photo).

sized pieces of the pulverized surface rock, cemented together in a glassy matrix. In addition to the layer of breccia, there is sometimes a lens-shaped layer of glass. Deeper down the original rock, while still intact, shows evidence of the violent event by being cracked and broken into rough blocks.

Mars is not a very active planet and we expect that the results of the period of major bombardment should still be in place under much of its surface. Subsequent to that period, volcanic activity has re-surfaced much of Mars, especially in the northern hemisphere and along the equator, so we see a generally younger, smooth surface. However, in the canyonlands, these old, shattered rocks ought to be exposed in the cliffs, providing a historical record of the bombardment and subsequent inundation by lava.

Above the shattered basement rock the expected layer of breccia may provide some water. This will probably be frozen water locked in the interstices of the fragmented rock and extracting it in usable quantities may be difficult. You can't just pile a bunch of rocks into an iron tub and heat them over an open wood fire. There isn't enough oxygen in the atmosphere for such a campfire, nor is there any wood lying around. It will probably be best to leave water extraction to large-scale engineering projects, such as those probably going on back at Mars Base.

Ridges and gullies

Camp 3, some ten miles from the canyon bottom, is planned for the layers of ancient breccia. Above it, the cliffs are probably steeper and the terrain is sharper. According to the images from Mars Global Surveyor, the upper cliff walls are made up of hundreds or thousands of layers of dark rock, probably basaltic lava flows. The form is that of sharp ridges separated by sharp gullies. While the ridges are mostly exposed hard rock, the gullies are filled with finer-grained material, such as sand and gravel. All is covered with a light sprinkling of wind-blown red dust, the whole effect bearing a remarkable similarity in appearance to the high peaks of the Himalayas, with the dust and sand substituting for the Himalayan snow.

Climbers to Camps 4 and 5 will have to choose between ascending the steep rocky slopes of the ridges or following the gentler gullies, which probably are prone to avalanches even more treacherous than snow avalanches on Earth. Slowly feeling your way among these dangerous alternatives, you will probably make slow progress. The layering should help. The Global Surveyor images showed layers that ranged from 10 feet or so to 150 feet in thickness. These are most probably similar to the layers of lava that are found in the valley walls of North America's Great Columbia Plateau. Skilled climbers can take advantage of the alternating cliffs and shelves found there, with the climbers on each pitch protected by those perched on the shelves with a secure belay. Progress will be slow, but you should be thrilled by the opportunity to explore this fascinating record of the volcanic paving of Mars' surface and by the knowledge that you are ascending one of the Solar System's highest and most spectacular canyon walls.

As you proceed you should make notes about the rocks you are on. You have a unique opportunity to settle a dispute among geologists about the cause of the layering. Most students of the orbital images believe that they are simply different lava flows, resulting from billions of years of sporadic, widespread volcanic activity. The light reflected from the rocks indicates that there is lots of the mineral pyroxene, which is common in volcanic rocks, and the darkness of the rocks is similar to the dark basaltic lava of the Earth and the Moon.

However, there are dissenting opinions. Some scientists have specu-lated that the layers are sedimentary in nature, resulting from the settling of sand and dust to the bottom of a giant ancient lake. When you

Layers in the cliffs of the Coprates
Canyon are probably the edges of
lava flows that have been exposed
by the formation of the canyon
(NASA).

Lava layers can be seen exposed in the cliffs of the Columbia Plateau on Earth. This valley was formed when a catastrophic flood poured over the hardened lava, exposing the layers (right and in the far distance) (author photo).

are right there, with your gloved hands on the rocks themselves, you probably will have no difficulty telling whether they are lava or sedimentary rocks, and the dispute will be solved. The answer will be tremendously important to the understanding of the geological history of Mars. Sampling the rocks as you go up is equivalent to a more difficult and much less adventurous task: drilling down from the surface through 20,000 feet and extracting cores.

The black band

The last few days of the ascent will possibly take you past a special mystery. Orbital images of Coprates from the Viking spacecraft showed one very prominent, thick layer that was much darker than the rest. Is this a thick layer of black basalt laid down by an unusual kind of widespread eruption? Or is it, as its unusual color suggests, a thick layer of black glassy material (called "mafic glass" by geologists), possibly extruded into the Martian rock layers? This famous "black band" can be traced over much of the south wall of Coprates but may not be present in the north wall. If it is glass, then it will be obvious as you climb up it. Its consistency will be quite different from the basaltic rock above and below it, being more fragile and crumbly.

Back to the surface

After ascending layer upon layer, fighting the steep pitches and avoiding the constant danger of massive avalanches, the climbing party will finally see the top of the cliff ahead. On the eighth day of exhausting toil, your final small group of "summiters" will find their way up the last section, possibly the steepest part, of the grandest canyon in the Solar System. Waiting there in the comfort of the Mars expedition vehicle, anxious to learn all the hard-won knowledge revealed by the climb, will be the rest of your party. There will be ample time during the long trek back to the Mars base for your group to share its experiences and to pool its wealth of hands-on data that will allow us Terrestrials to reconstruct the geological history of one of the great wonders of another world.

5

A polar crossing

Having climbed the highest mountains in the Solar System and explored its deepest canyon, are their any challenges left for you? Of course there are. On Earth, people are ingenious enough to think of plenty of astounding things to do: one can be the first to climb all 8,000-meter peaks, the first to circumnavigate Greenland, the first to sail alone around the world, and a seemingly endless list of other adventures. In this next Martian expedition, you will be accomplishing a valuable scientific mission as well as chalking up an important "first." But it is not a strenuous adventure involving athletic feats or physical danger. In fact, it is a comfortable, relaxed exploratory trip to a vast and mysterious land, the south polar cap of Mars.

Astronomers have known for centuries that Mars, like the Earth, has polar caps. Often these bright white features are the most conspicuous things seen when Mars is viewed by telescope from the Earth. Even a small telescope will show them to you. When Mars is at perihelion (the point in its orbit nearest the Sun) and at opposition (when Mars, as seen from the

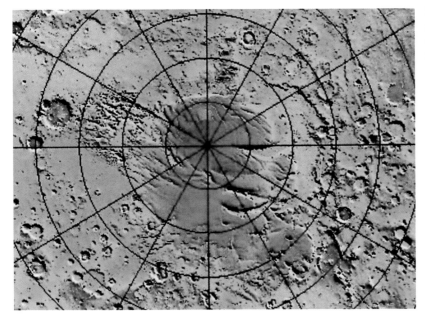

The south polar cap of Mars is shown here in a map based on elevations measured by the Mars Global Surveyor (NASA).

The edge of the polar ice shows layering, caused by seasonal and climatic changes (NASA).

Earth, is opposite the Sun), you can see the planet as a reddish-orange disk with a white circular patch contrasting starkly with its surroundings. This would be the south polar cap, which faces toward us when Mars is near perihelion and when its southern hemisphere is warmed by the summer sun. The northern cap is similar but is more difficult to see from Earth because it is tilted towards us when Mars is more distant.

Polar ice

Early astronomers recognized the Martian polar caps for what they are because they are round, they are centered at the two poles and they shrink in the summer. Winter at one of the poles is a cloudy, hazy season, with the entire area down to latitudes of 60° or so obscured. When the clouds clear, that whole area is covered in white. It is stark and bright and its edges are crisp. With the coming of spring, the snow or ice retreats slowly until by mid-summer only a small remnant remains. The retreat is not perfectly even; in some areas the white remains longer than in others.

One of these areas is a region at latitude 75° south that early astronomers named the "Mountains of Mitchell," assuming that snow remained there because of higher altitudes, just as our winter snows remain longer on the Rockies than in the plains on either side. If you look at a modern map of Mars, however, try as you will, you won't find the Mountains of Mitchell there. The Mariner and Viking spacecraft took excellent pictures of that area and turned up no mountains, just some craters, surrounded by an odd, layered and terraced countryside. The snow remains there for other reasons, perhaps because of the particular slant of the sunlight on the terrain. The Mountains of Mitchell were named for Ormsby Mitchell, a nineteenth-century American astronomer who has not been forgotten, just because these mountains turned out to be phantoms. A nice, large (60 miles across) meteorite crater to the north-west of the "mountains" is named for him.

But is the white stuff at the poles really snow? Until fairly recently most astronomers assumed that of course it is snow — nice, white frozen water, the kind that melts on Earth when the temperature gets above 32 °F. Percival Lowell even suggested that the molten caps provided liquid water for the rest of Mars, carried towards the centers of Martian civilization by his famous canals. As the twentieth century advanced, however, doubts arose.

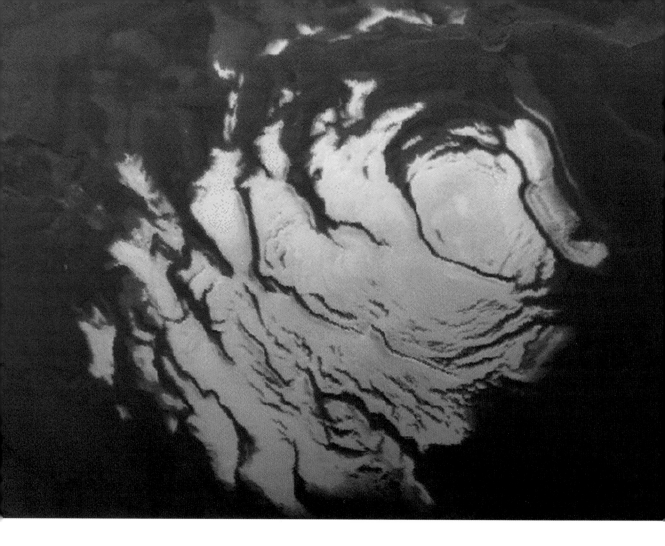

The polar ice has a roughly spiral shape, reflecting the direction of the prevailing winds (NASA).

It was discovered in the 1960s that Mars' atmosphere is very thin. The pressure at the surface is too low for water to remain liquid very long. Instead, water ice, if heated, would evaporate (actually, the correct technical term is "sublime," meaning that the ice goes directly into the gaseous phase, skipping the liquid form). Furthermore, the temperatures measured for the polar regions were found to be too cold for water ice to melt or sublime to a large extent. At the south pole, the balmiest summer day doesn't see the temperature rise much above −175 °F. Small amounts of water vapor are seen above the south polar regions, but it must be higher in the atmosphere where it is warmer.

The north pole is different. Even though its summer season occurs when Mars is near its farthest point from the Sun, temperatures at that pole can warm up to −40 °F, a reasonable temperature for anyone used to a winter in Siberia or Minnesota. In Mars' thin atmosphere, such a day is warm enough for water ice to partially sublime and, indeed, astronomers detect an increase in the amount of water vapor over the North Pole in mid-summer. It is nowhere nearly enough, however, to explain the bulk of the shrinking polar cap. Something else must be there, something with an even colder freezing point than water.

This puzzle was solved in 1977 when it was suggested that dry ice (CO_2: carbon dioxide) was the logical material. Temperatures at the poles are cold enough in the winter for CO_2 to precipitate out of the atmosphere to form frost. The warmer polar temperatures in the summer are just right for dry ice to sublime (its frost-point in Mars' atmosphere is about $-200\,°F$). Thus the clouds that shroud the poles in winter are mostly the kind of dry ice mists that people use in theaters and punch bowls to create a mysterious effect, and the retreat of the snow is the return of the dry ice to the Martian atmosphere as a warmer, transparent gas.

North vs south

The Earth's two polar regions are quite different from each other. The north pole is located in an ice-covered ocean while in the south there's an ice-covered continent. We shouldn't be too surprised, perhaps, by the fact that Mars' two poles also differ. The north pole has a permanent cap that consists primarily of a layer of water ice. Its surroundings are relatively smooth and sand dunes are scattered about. The residual ice cap is nearly centered on the pole and it shows a clear spiral pattern, caused by winds.

Mars' south pole is more of a mystery. Its temperature is never high enough to let us know whether water ice is present at all. We see only the carbon dioxide vapor as the dry ice sublimes in the southern summer. The remnant cap may or may not be made of water ice. The best way to find out, probably, is to go there and sample it.

The south pole has other peculiarities. For instance, the permanent ice cap is not centered on the pole itself, but is badly off-center. By mid-summer the seasonal CO_2 cap has shrunk to the point that leaves the true pole at the edge of the cap, with the ice extending 200 miles northward on one side. This asymmetry may explain why its outline, while some-what segmented at its borders, is not spiral like the north pole's.

The polar trek

The mysteries of the south pole are legion. Does the largely seasonal dry ice cover a hidden store of water ice? What is the nature of the strange

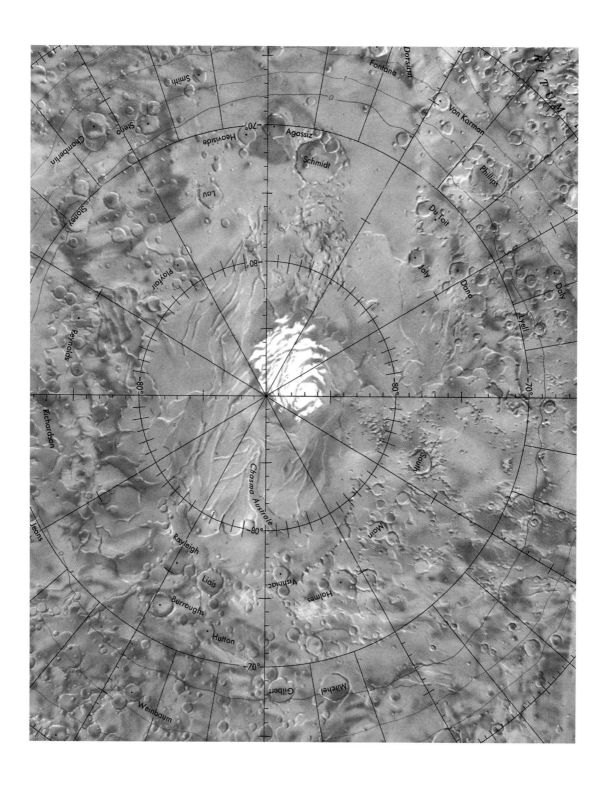

layers that are seen in the polar hills and valleys? What caused the giant polar canyon? Why is the permanent cap so far off-center? What are the black spots discovered by the Mars Global Surveyor? What made the weird scalloped surface on the permanent ice? And can the polar deposits reveal, layer by layer, the history of Mars' weather and its climate?

To answer these questions and others, your plan is to mount an expedition to the pole and to cross it from one side to the other, studying the land and ice and collecting samples as you go. The trip will take 30 days, averaging only about 35 miles per day, allowing plenty of time for exploration. Transportation will be by a giant track-mounted snowmobile that not only provides the comforts of home (heat, air, food) but also has a small laboratory for the instant study of collected samples.

The plan is to start the trek at the latitude where the winter dry ice layer extends to near its maximum, at latitude 75°S, from there to proceed directly south to the pole, and then to continue up and over the permanent cap, stopping at latitude 75°S on the other side. The trip is planned for mid-summer to allow your crew to study the rocks and deposits, as well as the mysterious permanent cap, without having to dig down through two or three feet of dry ice.

[opposite] The route of the polar expedition across Mars' south pole can be followed on this map from the bottom, near crater Mitchell, to the top, near crater Agassiz (from a USGS map).

The edge

The ice and snow of the poles has had a most definite effect on the land. All around it at modest latitudes, the southern Martian surface is an ancient cratered terrain, looking remarkably like the Moon. Little has changed it over the last 3 or 4 billion years, other than wind erosion and dust storms. Almost everywhere you look are impact craters, some as large as 100 miles across or more. But in the polar regions the craters are almost gone, covered with a smooth, patterned landscape that is unlike anything found elsewhere on Mars, except at the north pole.

The polar trek begins in the midst of the ancient cratered terrain where the effects of the polar ice begin. You start out at latitude 75°S, longitude 270°W, a position near a small 12-mile crater in a smooth plain. The land is covered with red soil, scattered black rocks, and a sprinkling of sand and dust. The horizon is broken only by some distant low hills. The sun is fairly bright, for Mars, but it fails to warm you. The noon-time temperature is a chilly −175 °F.

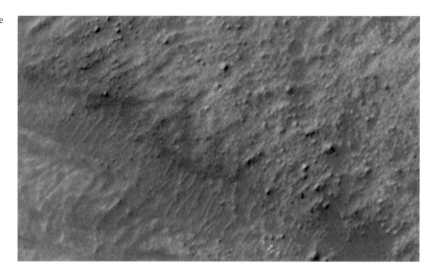

The surface of Mars near the pole is mostly sand and dust, covered with dark boulders (NASA).

This is the land of the midnight sun on Mars. You're 15° from the pole and Mars' axis is tilted by 24°, almost the same as the Earth's. Therefore, at this summer season the Sun never sets, but arcs across the sky in a wide circle, providing plenty of light to explore during a long work day. Near midnight, when the Sun is lowest and most dimmed by haze and dust, you may be able to see a bright star not far from the Sun. Look for it with binoculars. This will be Earth, a thin crescent, passing swiftly between your adopted planet and the Sun as it follows its annual circuit of the inner Solar System.

The first day will take you through transition country, passing from relatively undisturbed old terrain to land that shows the effects of winter snows. The task is to look for the effects and study how they mount as the expedition moves southward. How sharp is the edge? What are the effects that are most conspicuous on the ground? From orbital images it is not possible to discern more than a large-scale gradual smoothing of the land, but there, standing on that land, you will be able to seek and discover many new things.

Vishniac

On the second and third days, the route will divert a little from its straight-south direction to visit the rim of a large impact crater, some 80 miles in diameter, which lies a few miles to the west. Named Vishniac after Wolf Vishniac, a twentieth-century American microbiologist, this crater is one of the largest found this close to the pole. This is a good place to study the effect of winter frost. As spring approaches, we often see frost outlining the craters near the polar edge and the frost often stays longer just inside the northern crater rims, where it is protected from the mounting Sun. Also, in years with strong winds, the east rims retain winter frost longer than the west, where the prevailing winds help to remove it. You'll be visiting the east rim of Vishniac and can do a thorough ground-study of the effects of this seasonal phenomenon.

Layers of history

Past Vishniac the route proceeds south for two days through a smooth, featureless plain. At least that's what orbiter maps show. It's certainly not featureless for your crew, who will see and explore it, studying the effects of the alternating seasons.

But on the sixth day there is something special to be encountered. Near latitude 80°S begins the famous "layered terrain" that covers most of the exposed part of the polar regions. Valleys, hills, low ridges and scattered mesas reveal that the polar lands are layer upon layer of flat deposits that alternate between bright and dark. The hills are gentle and are separated by widely-spread valleys that are broad and a few hundred to a thousand or so feet deep. The valley walls are striped by the layers of deposits that are 20 to 100 feet thick. Terrestrial geologists believe that these layers are made up of sediments of wind-blown sand and dust (including the dust from volcanic eruptions) that record the history of Martian seasons. Data from the crater density suggest that the layered terrain is about 100 million years old. The changes in the history of Mars, which seem to be so important and sometimes so surprising, are laid out like the pages of a book, telling us about the periods of warmth, of deep cold, of volcanic eruptions, of dust storms and probably much more. You'll be moving across this layered history for many more days, time enough to begin to translate the stories that the layers tell.

The Great Southern Canyon

After mounting the layered plateau, your path crosses it for a few hours and then comes to the top of a steep cliff. From there, spread out before you to the south, is a giant canyon, known officially as "Chasma Australe." The smooth-floored canyon is 50 miles wide. It is some 300 miles long, extending from the poleward direction to about 80° latitude, apparently emptying onto the plains just beyond the layered terrain.

How was it formed? By liquid water? Not likely. The temperatures here, at least now, are far too cold. By the glacial action of moving ice? Probably not. To move, glaciers need to be near the melting point so that the ice is elastic and so that liquid water can help "grease" the movement. At $-175\,°F$ even carbon dioxide ice is unlikely to be able to flow.

Perhaps the wind did it. We do see sand dunes in the polar regions and the ice patterns point to the importance of wind. Your crew will want to spend plenty of time taking measurements and samples so that the riddle of the remarkable Great Southern Canyon can be solved.

Ridges and valleys

For three days following your time in the canyon, the expedition will traverse a frigid and bumpy land, where for more than half of each Martian year the ground is covered with dry ice. The best maps of this area show a succession of ridges and valleys, with the layering exposed on the valley walls, ripe for study. The best maps available now, however, show little detail. This is largely a land of mystery, the kind of place that can only be understood by adventurous explorers on the ground. Perhaps it will be monotonous. Certainly the pioneers of polar exploration on the Earth, as they crossed the northern ice islands or pushed their sleds across Antarctica, had days when the scenery seemed monotonous enough. But Mars may be different. Having chosen to cross this terrain in summer when the ice is gone, you may be overwhelmed by the complexity of it. Geologists and astronomers at home will be anxious to hear your story.

At the pole

Finally on the twelfth day you reach the pole. Standing on that spot, no matter which way you turn you're looking north. As the Sun makes its daily circuit in the sky, the shadows in the rocks and cliffs do their daily dance, leaning one way and then the other. Conditions are just right for a few days of exploring, inspecting and collecting.

The most important question to answer is the long-standing puzzle of the permanent ice cap. Its edge is somewhere here, very close to the pole. What kinds of ice make up the south permanent cap? Is it dry ice underlain by a permanent layer of water ice? If so, how thick is the water ice? Does it contain within it dust and debris from long ago that might add to our knowledge of Martian history? These and other questions will occupy your mind as you and your crew walk up to the edge of the ice, ready to take samples, cores, and lots and lots of pictures.

An imaginary view of the Martian south pole, with a dusty, pink sky and the pinkish ice, partially covered by wind-blown sand and dust.

Crossing the ice

It's time to drive your giant snowmobile onto its first snow. We don't know how difficult it might be to mount the permanent ice cap. Perhaps you will just drive from dry land onto ground that is covered with an increasingly thick snow layer, as if you were driving north in a New England winter. Perhaps the snow on either side of your vehicle will be dirty, as on a Boston street in December. Or perhaps not. Unlike terrestrial snow, this stuff is mostly (possibly entirely) *dry* ice. There's no melting and no slush.

It will take over a week to traverse the ice cap. Much of that time will be spent at strategic stops for sampling and geological exploration. But

The edge of a terrestrial glacier is often dirty and rock-covered, as the ice abrades the mountain's rock as the glacier flows downwards (author photo).

there's no point in sending humans to such a remarkable place unless they can take advantage of it in some uniquely human way. That's why some of your crew will have brought their skis. There probably aren't really good slopes for downhill skiing, but the nordic skiing might be excellent. With no night to stop you, many hours of cross-country traversing await you each day. And for the non-Norwegians on the expedition, being towed on skis behind the snowmobile might be adventure enough. Of course, you will have special solid-state Martian polar skis on board, the kind developed for use on carbon dioxide snow (or perhaps some extremely cold water ice).

The days on the ice cap may provide clues to another strange mystery. The south polar permanent cap, unlike its northern sister, is cockeyed. Why isn't it centered on the pole? Why are its edges arranged in giant arcs, open to the south, while the northern cap has a more perfect spiral pattern? These facts are possibly related to the working of Martian winds and to the violent dust storms that frequent the southern regions in summer. In fact, your expedition may already have experienced one of these storms, when the red sand and dust is stirred up to great heights until the entire landscape is blotted out. Sunlight, otherwise so reliable on sunny Mars, is obscured so much that day turns almost to night. The headlights will have to go on in the snowmobile and extra-vehicular travel will have to wait for the storm to abate.

Honeycomb hills

At the end of the third week, your expedition will reach the far side of the ice cap. We don't know what the edge is like. Perhaps your vehicle will just drive down off the ice and go "bump" onto land, like the snowmobiles leaving the Columbia Ice Field in the Canadian Rockies. Or it may be more gradual. In any case, the return to the red Martian landscape will provide new opportunities to explore the nature of the layered terrain that surrounds the cap.

Rather than continuing in a straight line along the 90th W parallel, the plan is to divert to longitude 65°W to allow the exploration of some intriguing landforms there. At a latitude of 81°S you will encounter some busy ridged terrain. At the end of the twentieth century, Mars Global Surveyor obtained images of this weird country, in which there are long, straight ridges that wall in large rectangular, closed valleys. What caused these strange geometrical shapes is a mystery. And added to that puzzle is the fact that sprinkled over the bottoms of some of these strange valleys are even stranger black spots, looking like some kind of giant pox.

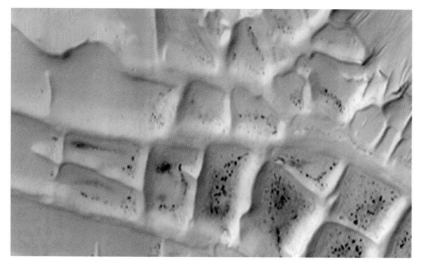

Mars' honeycomb hills, sometimes called "Inca City," lie at the edge of the polar ice (NASA).

The spots were free of snow, although frost thinly covered the rest of the area when Surveyor saw them (in late spring), so whatever they are they warm up faster than normal.

Martian geologists believe that this part of the south polar region may contain clues to a very complex history. It has been suggested that the strange rectangles may be wind-sculpted ridges that are remnants of an ancient terrain, older than the polar layers, and therefore it will be important to gather many samples and to explore these enigmatic features thoroughly. They may open the book to even older Martian history than is written in the layers of the polar surface.

Back to the ancient craters

The last few days of the polar trek will be spent passing through the ridged geography onto the ancient southern plains. You will move from recent deposits laid down and disturbed by the annual snowfall and the polar winds onto a bare, rocky cratered surface, much like that from which you left a terrestrial month before. By diverting slightly left of north from the rectangular valleys, the expedition can head towards an exceptionally large crater, the giant impact scar named Schmidt. Its name honors two scientists of the past, Johann Schmidt, a nineteenth-century German astronomer, and Otto Schmidt, a twentieth-century Russian geophysicist.

The crater Schmidt is about 150 miles across. As you near the southern rim, you will have more of an impression of being at the top of a scarp rather than at the edge of a crater. The far side will be lost to view on the hilly northern horizon. Standing on the rim, looking down, you may be reminded of the view seen from the rim of the Mogollan Plateau in northern Arizona, where a steep descent from a high mesa leads down to a flat, dry desert.

Schmidt is an old crater, as can be seen by the many more recent craters that lie on its floor and across its northern rim. Being so near the pole, it is subject to some episodes of winter frost, so there ought to be some features of erosion worth exploring. But that will be the last science to be done on this expedition. It is time to return home to Mars base. With the lockers full of rock and ice samples, the notebooks full of maps and the electronic cameras full of images, the snowmobile must end its polar trek and bring its treasures to the many people awaiting its return.

6

The other Alps – climbing Mt. Blanc

For mountain lovers the Alps are the jewels of Europe and many people believe that Mt. Blanc, the highest in the range, is the crown jewel of the Alps. At 15,781 feet above sea level, Mt. Blanc's white and gleaming bulk dominates the western end of the mountain chain. The excellent hard rock of the Massif has led to the existence of many very sharp, steep peaks in the area and to many hair-raising rock-climbing routes. On any sunny summer weekend over 5,000 people can be found hiking, clambering, climbing, rappelling, and glissading on the slopes of Mt. Blanc and its companion peaks. The heights are reached by trails, mountain trains, aerial tramways, and occasionally by parachutes and paragliders.

By contrast, the other Mt. Blanc has no crowds, no trails or tramways, no snow and no air. Like its terrestrial namesake the lunar Mt. Blanc is the highest peak in its range, named (naturally) the Alps. But that is one of the very few similarities between the two Mts. Blanc.

The terrestrial Mt. Blanc is the highest peak of the Alps on Earth (Gordon and Sara Hodge photo).

The lunar Mt. Blanc lies in the lunar Alps (Montes Alpes) east of Mare Imbrium. The Alpine Valley cuts through the mountain range (from a USGS map).

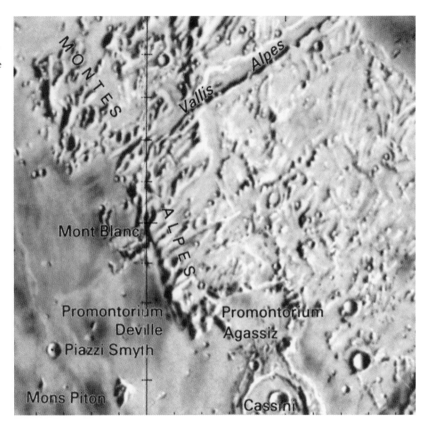

Even their beginnings were different. The Alps of Europe formed gradually over millions of years, as slowly moving sections of the Earth's crust pushed together, squeezing the land to form a giant arc of up-thrust mountains. The lunar Alps, in contrast, were formed in an instant, when a giant asteroid almost 100 miles in diameter collided with the Moon about 4 billion years ago. The collision produced a mighty explosion that melted and shattered large amounts of the Moon's surface, making a huge crater some 750 miles in diameter. This crater was eventually entirely filled by basaltic lava, forming the dark, circular basin known as Mare Imbrium (the "Sea of Rains"). Of course, the Moon has virtually no atmosphere and no water, so the feature was sadly mis-named, but it does vaguely resemble a sea, with its comparatively smooth surface, formed by a now-frozen lake of lava rather than water.

Impact craters have raised rims and steep inner walls. The giant Imbrium crater was no exception. After the explosive collision, when the fragments, rocks and dust had fallen back to the surface, the giant hole was surrounded by at least two concentric circular rings of mountains that made up its "rim." The inner, lower one was eventually almost entirely engulfed in the lava that filled the basin (see Chapter 7), but the outer one remains today as a circular series of arc-shaped mountain ranges. In places these mountains rise more than 10,000 feet above the lunar plains. Generally, their inner walls are steep and well-defined, with the hinterlands being more broken and sketchy as the elevation gradually decreases away from the impact site.

Early European mappers of the Moon divided this ring of mountains into several ranges and named them after European mountains: the Alps, the Juras, the Caucasus, the Carpathians and the Apennines. The rim is incomplete in several places around the basin, making this separation into distinct segments a natural one.

Although everyone has their favorite mountains, to my mind the Alps and the Apennines are the most spectacular and the most geologically interesting of all lunar mountain ranges. Many years ago the Apollo 15 astronauts visited the Apennines, putting their spacecraft carefully down in a valley beneath the high slopes of Mt. Hadley. But until now the Alps have not been explored on foot nor has their highest peak, Mt. Blanc, been climbed.

Mt. Hadley shows the typical lunar mountains' barren slopes. It was photographed by the Apollo 15 astronauts (NASA).

Plato — western sentinel of the lunar Alps

The terrestrial Alps extend in a thick giant arc from eastern France through Switzerland, Italy and Germany to the far western part of Austria. The lunar Alps extend in a thick giant arc from Plato to Cassini. Superimposed on the northwestern tip of the Alps, Plato is a beautiful impact crater, easily seen from the Earth with a small telescope. It is particularly conspicuous and interesting because of its smooth, almost featureless, dark floor, made of lava that completely fills its basin. About 60 miles across, Plato's floor was paved by a series of lavas of low viscosity some time after the impact event. The slightly mottled surface suggests a sequence of different lava flows. Five or six more recently formed small craters, a mile or less in diameter, disturb the flat floor of Plato, and there is a curious, inconspicuous dome near the eastern "shoreline." This mysterious feature may just be a hill or ridge that wasn't completely

The ancient crater Plato lies at the northern end of the lunar Alps (NASA).

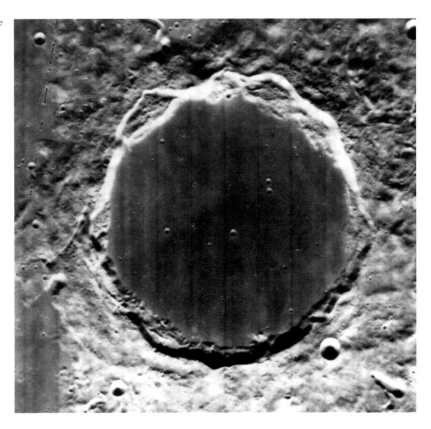

covered when Plato's floor was flooded, but the crack in its summit suggests that it might instead be a very small shield volcano.

There are additional volcanic features outside Plato's walls. The crater is surrounded by a beautifully patterned blanket of ejecta: ridges and valleys radiating outward and formed by the lunar material that was violently thrown out by the impact. Amidst this woven skirt of jumbled rocks are several lava river valleys, formed by lava that oozed up from fractures and then flowed in a meandering path down the outer slopes of the crater. The two most conspicuous valleys, called "lava channels" by volcanologists, are on the east and west slopes of Plato, each starting from small source craters or vents about 20 miles out from the main crater's rim. The western channel is double, nestled in an older, wider valley about a mile in width. There is a much narrower but longer channel that is just barely visible on Lunar Orbiter images. Apparently at least two volcanic episodes were involved at different times, both long ago.

The Alpine Valley

To the east of Plato the Alps are a rugged, wide range of rolling hills, punctuated by many higher peaks. About 120 miles along the range from Plato, however, is a famous feature, the great Alpine Valley, which abruptly cuts through the mountain chain. Nearly 120 miles long and straight as an arrow, the Alpine Valley is a spectacular feature as seen in a moderate-sized telescope and a favorite sight for amateur astronomers. It looks for all the world like a giant canal, dug to connect the Sea of Rains (Mare Imbrium) with the Sea of Cold (Mare Frigoris). The valley's nearly flat floor is covered with smooth, dark mare material, indicating that liquid lava did once flow through the valleys, only to freeze solid as basaltic rock some 3 billion years ago.

The Alpine Valley averages about 5 miles in width. Were you to stand in the middle of its flat floor, the sensation would be somewhat like standing in a terrestrial river valley that cuts through a modest and ancient mountain range. The walls of the valley are fairly steep for lunar cliffs, which are seldom as sharply defined as photographs make them look. The linear form of the Alpine Valley and its location between two giant impact sites is pretty clear evidence that it was formed as a crack in the lunar crust (what geologists call a graben) that opened up as a result of the giant impacts and was later flooded by the lava from the lava seas.

The Alpine Valley is the small, straight feature above and to the left of the center of this ground-based photo. It cuts through the lunar Alps, connecting the Sea of Rains (Mare Imbrium) with the Sea of Cold (Mare Frigoris), seen here to the right of the valley (Mt. Wilson Observatory).

A close inspection of the Alpine Valley shows that it is double, like Plato's smaller lava channel. Meandering down the middle of the valley floor is a narrow stream bed. This is probably a channel that formed later during a subsequent volcanic period. Some lunar scientists have even identified a possible volcanic source crater near the valley's southwest end. Furthermore, there are several small mountains on either side of the valley that look suspiciously like volcanoes; some are high, rounded peaks with small craters at the top and one is a beautiful, symmetrical ring that strongly resembles tephra volcanoes on Earth, which are made of ash-like ejected rocks. There is not universal agreement about these features and we need to go there to understand completely the remarkable sequence of events that formed the great Alpine Valley.

A crater curiosity

Between the giant crater Plato and the Alpine Valley, the Alps are mainly blocky, irregular hills and peaks, interrupted here and there by small impact craters. For example the crater Trouvelot lies just south of the valley near its eastern terminus. Named for the nineteenth-century French astronomer Etienne Trouvelot, this fresh-looking crater is about 5 miles across and looks like a perfectly normal lunar impact crater.

Another fresh-looking crater to the north, however, presents a curious puzzle. Slightly larger than Trouvelot, this crater has yet to be given a proper name. It is currently designated "Alpes A," the first in a series of

The mysterious Alpes A (NASA).

nameless craters that include other letters of the alphabet, up to at least "Alpes W." But it is Alpes A that deserves our special attention.

If not overwhelmed by more recent cratering or by mare lava flows, most lunar impact craters, if they are fresh enough, show a surrounding blanket of features caused by the materials ejected by the impact. Among these features are long rays of spattered dirt and rocks and small, elongated secondary craters created by the larger chunks of moon rock hurtled out and away from the crater. The secondary craters are normally elongated in a radial direction, as the blocks of rock dig them out in a glancing impact. The giant crater Copernicus, for example, located near the center of the face of the Moon, is surrounded by a glorious set of bright rays and secondary craters (Chapter 8).

A close look at Alpes A, however, shows a peculiar difference. Its outer rim and immediate vicinity is pocked with small craters, rather more than might be expected for an impact crater only 6 miles across. But the really curious thing about this blanket of craterlets is their shape. They are all round, not elongated. If they are secondary craters, how in the world could they have the shape of perfectly round little pits? Ejected blocks of rock hitting the ground close to the impact would normally plow into the soil, making a groove-like elliptical crater. The perfectly circular pits of Alpes A suggest that all the ejecta from this curious crater were thrown almost straight up, so that they would sprinkle down nearly vertically. How could that happen? No one knows.

There is a different, but controversial, explanation for Alpes A's puzzling pits. Perhaps Alpes A is not an impact crater at all, but a volcanic crater in disguise. Volcanic eruptions can throw chunks out nearly vertically and they then rain down as bombs, making quite circular pits. Alpes A may be a dead volcano, the secrets of its explosive past almost hidden in the rugged terrain of the lunar Alps.

Two capes

Continuing the survey of the mountain range to the southeast, we come at last to the high peaks of the range. Because the Alps were formed as the high rim of a giant crater, the highest peaks are at the inside of the arc, where the rim was steepest. The walls of the basin now are expressed as high cliffs that rise abruptly from the crater floor, which is now the lava-filled Mare Imbrium.

As explorers there on the Moon you will get a true impression of the steepness of those cliffs and of the abruptness with which they rise. Through Earth-based telescopes the lunar mountains appear to be rugged, indeed, with near-vertical slopes. That is because it is easiest to make out the lunar features when they are experiencing lunar dawn or lunar dusk, when the shadows greatly exaggerate the surface relief. As the Apollo astronauts found, the true slopes are more gentle. The billions of years of gradual erosion by small meteorite impacts have softened the contours of even the highest mountains.

There are three high peaks in the Alps that have been named. Two of these are at the southeast end of the range, rising above the sea like the prow of an ancient ship. True to the marine analogy, they are named "capes" ("promontorium" in Latin). The southernmost, Cape Agassiz, towers above the lava sea even more prominently than Cape Horn rises above the seas beneath the tip of South America on Earth. South of Cape Horn beyond the stormy seas, is the ice-covered continent of Antarctica, while south of Cape Agassiz, across a bay in the Sea of Rains, is a more naturally lunar feature, a large crater named Cassini.

Cape Agassiz was named after a prominent Swiss naturalist, Louis Agassiz, who graced the nineteenth century with his prolific studies and writings. Another nineteenth-century scientist, Sainte-Claire Deville, a French geologist, is memorialized by another high Alpine peak named Cape Deville, which is 15 miles northwest of Cape Agassiz. About 30 miles farther on in this direction is the third named Alpine feature, Mt. Blanc, set back somewhat from the front range, but massive and high, with its highest peak nearly 12,000 feet above lunar "sea level." A planned expedition to climb this magnificent mountain forms the rest of this chapter.

Across the sea

We assume that your trip to the Alps begins at a lunar base located to the southwest, somewhere near the center of Mare Imbrium. Although the distance is a modest 400 miles or so and the terrain is flat, you should plan on spending a full three days driving to the base of Mt. Blanc, because there are a few quite interesting things to see and explore along the way. Some of these are isolated peaks and small ranges that are all that remain of inner ring structures of the Imbrium basin.

The Spitzbergen Mountains rise up above a bay called Sinus Lunicus, next to Mare Imbrium, the Sea of Rains (from a USGS map).

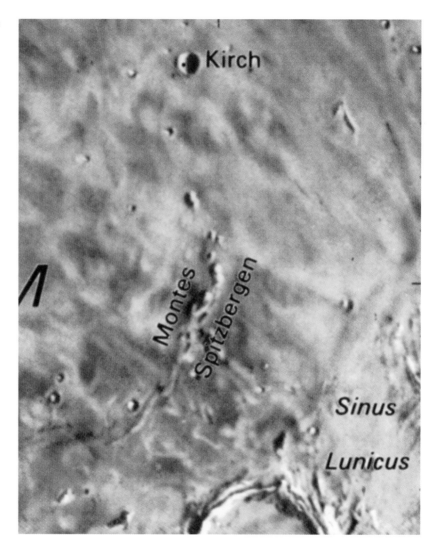

The first stop is at the Spitzbergen Mountains, reached at the end of a long first day's drive. This chain of modest peaks is about 40 miles long and its high points are close to 5,000 feet above the plain. The range is similar in height and shape to New England's White Mountains, but there are few other similarities. The Spitzbergens are bare rock, formed as part of an inner ring by the Imbrium impact and flooded up to its ears by the lava that filled the basin. What you see when you stop there is like the tip of an iceberg, with most of the mountains hidden in the lava beneath your feet.

Mary Blagg, the prominent British selenographer, named these mountains after the barren and rocky Spitzbergen, an Arctic island that lies east of northern Greenland. In spite of its remote and forbidding nature, Spitzbergen is now visited by cruise ships, full of people like you who thirst for adventure and the thrill of exploring the exotic.

There is a narrow valley that cuts right through the lunar Spitzbergen Mountains and that is a good place for your expedition to spend the night. You aren't seeking the shelter of the valley the way you might at the terrestrial Spitzbergen, where it could protect you from violent arctic storms. Instead this place is chosen because, with the mountains close at hand on either side, it is a convenient base for a morning's geological exploration. We have few giant multi-ringed impact structures left on Earth, so our knowledge of their structure is largely hypothetical. Simple, small impact craters are well-studied on Earth, but only on the Moon and on certain other Solar System bodies can the complex multi-ringed basins be subject to the geologist's hammer and practiced eye.

The second day's drive takes you north across the barren plains of the eastern Imbrium seascapes to the foot of a famous isolated peak called Mt. Piton. The stop here for the night is intended mostly as a reconnaissance in preparation for further mountain-climbing adventures,

The terrestrial Spitzbergen, an island in the north Atlantic (Dale Smith photo).

described in Chapter 7. The following morning involves a relatively short drive, about 60 miles, to the shoreline of the Sea of Rains and the base of Mt. Blanc.

Climbing Mt. Blanc

Arriving at the mountain in mid-morning will leave you the rest of the day to study the slopes and plan your climb. Of course, "mid-morning" and "day" mean something only to your internal clock in this case. Lunar morning will have occurred a terrestrial week earlier and the entire expedition to Mt. Blanc will be made during continuous bright, sunny lunar daytime. The Moon's rotation period with respect to the Sun is nearly 30 Earth days (hence the word "month") and so any spot on the Moon experiences about two Earth weeks of sunlight followed by two Earth weeks of darkness. Of course, you will have arranged the Mt. Blanc expedition to happen during the lunar daytime. And it's a good thing, as there is much to see and do on the way up to the summit of the Alps' highest mountain.

Mountain climbing on the Moon presents a few special problems that the Apollo astronauts did not have to worry about. Their various trips away from their spacecraft (EVAs, "extra-vehicular activities") were timed so that they would always be able to return for meals and rest periods. Eating lunch away from an air-filled spacecraft is no picnic. Locked as you will be in a leak-proof space suit, there is no chance (nor will there be any desire) for you to stop half-way up the mountain, take off your helmet and eat your sandwich. Instead, all of the food for the ascent is stored inside your suit and made available by controls on the outside of the suit, manipulated by your gloved hands.

The awkwardness of these arrangements argues for a fairly rapid ascent and return, so the plan is to get a very early start, at perhaps 3 AM your time, with a return to the base timed for evening. In this respect, the climb is not different from climbs in the terrestrial Alps, where similarly-high mountains are climbed in a single, exhausting day.

Compared with Earth-bound climbers, you have the distinct advantage of the low lunar gravity, making your body light and your steps bouncy. But you also have the disadvantage of wearing a bulky and heavy space suit, so that progress up the mountain will not be the hop-and-jump frolic that would be the case on flatter lunar terrain. It will take all of the muscle-tone and determination of a mountain climb back home.

Although Mt. Blanc looks steep when viewed from the Earth at its sunrise, your view from the base shows more gentle slopes, appearing more like the high, barren slopes of the central Rockies. There probably will not be any technical rock climbing involved, but rather just a long, slogging trip up a rocky, dusty series of ridges and valleys. The western side of the mountain is the steepest but the most easily accessed from the Imbrium plains, so that will be your route. The direct ascent from base to summit involves a straight-line distance of only about 5 miles, but you will want to pick your way along a more gradual route, switchbacking across the slopes and doubling or tripling the distance covered.

As you ascend there will be some important research to be done, involving collecting rock samples and taking lots of photographs. Our understanding of the Alps is still somewhat uncertain. We know that they were formed by the giant Imbrium impact, but we are fairly certain that they are not a crater rim like those formed by smaller impacts. For craters 50 miles across or less, there is a rim of up-thrusted and overturned rock, folded over like a blanket. These features are beautifully displayed at several terrestrial meteorite craters, such as the Barringer crater in Arizona or the Wolfe Creek crater in Australia. Most geologists believe that the Imbrium event was much too gigantic to have formed a single overturned lip of this nature.

Instead, the mountains forming Imbrium's rugged boundary may have been caused by slumping. The highly disturbed outer, uplifted surroundings of the crater may have relaxed after the initial violent upheaval, forming large concentric cracks in the crust, and leaving a series of two or perhaps three hilly rings around the crater (Chapter 7). Most of the inner rings were later inundated by the lava lake, but the outer rings, probably located quite a distance beyond the actual edge of the crater, remain. As you make your way up Mt. Blanc, you will want to search for evidence of these possibilities. Are there signs of overturned layers as in crater rims? Is there, instead, evidence of layers that have been tilted away from or towards the crater? Are there signs of slumping? Are the rocks mostly breccia (impact glass enclosing broken fragments of the impacted rock) or are they comparatively undisturbed layers of the ancient lunar crust? These and other questions will motivate the frequent stops for photos and rock sampling that will punctuate your climb.

The view from the summit should be spectacular. Any real mountain, no matter how desolate, can provide a special thrill from its top. You will be able to gaze southeast along the high ridge of the main Alps, with Cape Deville and Cape Agassiz forming prominences on the horizon. Off to the southwest will lie the flat, bleak floor of the Sea of Rains. To the northwest is a deep bay and beyond it the continuation of this great range. Northeast are the rugged hills of the outer Alps, pocked by small craters and the ridges and valleys of ejecta.

The trip back to your vehicle will be much quicker than the climb up. You'll no longer need to stop for scientific reasons or to rest. On the steeper slopes, especially if you find them to have a loose layer of dust and dirt, you might even be tempted to glissade a while, sliding down this make-shift substitute for snow. You had better do this with great care, however, as you won't want to fall and tear open your space suit, one of the special dangers of lunar alpinism.

As on its terrestrial namesake, the lunar Mt. Blanc looks passive and eternal, but it should be given due respect for the hazards that might lie hidden. By dinner time, though, any dangers will have been forgotten as you and your companions gather in your vehicle to celebrate a glorious first ascent, made more than 200 years after men first conquered Earth's gleaming white peak that mountain-lovers call the crown jewel of the Alps.

7

Pico Peak — monadnock of the Moon

There is a lonely mountain in the lunar Sea of Rains that is officially named "Mons Pico," which, when you translate Mons from the Latin, is "Pico Peak," which, when you translate Pico from the Spanish is "Peak Peak." With such emphatic redundancy you might expect that Pico should be a spectacular peak indeed, and if you did you would be agreeing with centuries of observers of the Moon. At lunar dawn or sunset, Pico Peak casts a long, sharp shadow that has evoked images of a steep and forbidding rock spire, worthy of world-class rock climbers. Is it possible to climb it at all? We will see.

Monadnocks on the Moon

Pico, regardless of its difficulty as a mountaineering goal, is of high interest as a scientific goal. It can be thought of as an example of what on Earth is called a monadnock. Named after a small, isolated mountain in southern New Hampshire in the US, monadnocks refer geologically to

Early telescopic views suggested that Pico would be a precipitous tower as seen from the Moon's surface (from a textbook published in 1882).

Mt. Monadnock is the isolated peak in New England that lent its name to a class of land forms (author photo).

non-volcanic mountains that occur alone in a landscape, separated from any mountain range. They are unusual, as mountains are usually formed on a large scale by large-scale forces, producing long chains of related peaks. On Earth monadnocks can have a variety of explanations, but this variety does not include the process that formed Pico Peak.

Monadnocks on the Moon are rare, as on Earth, and they are limited to locations in the great volcanic seas. In Mare Imbrium, the Sea of Rains, for example, there are a few others besides Pico, such as Piton Peak and Mt. La Hire. They rise above the flat, dusty plains and seem unrelated to their surroundings. How can the Moon have isolated, sporadic mountains like this? Are they lone volcanoes that have sprung up from random holes in the crust? Probably not, as their form is not symmetrical, no flows of lava can be seen mantling their sides and spilling onto the surroundings and no craters lie at their summits. So how otherwise does a lone peak like Pico Peak form?

The goal of this chapter is to plan a lunar expedition to answer this question. The trip will begin at the same lunar base as you used for the expedition to Mt. Blanc and you will have an opportunity to visit some related features near the northern shores of the Sea of Rains along the way. But first it will be wise to examine the current theories about the origin of these lunar monadnocks so that you will know what to look for when you are climbing Pico Peak. Not only will you have a chance to climb a mountain that has stimulated the imaginations of centuries of lunar observers, but you may be able to settle a question that has puzzled the most perceptive of lunar theorists.

More than one

Clues to the solution of the mystery can be found both on the Moon and back here on the Earth. The first place we'll look for them is at a lunar feature that was a tantalizing target for telescopes before the space age gave us proper pictures of it. Mare Orientale, the Eastern Sea, barely can be glimpsed from the Earth at the very edge of the Moon. Not until the Lunar Orbiters mapped the whole lunar surface did we get a good idea of the full nature of this beautiful circular basin and then we found a remarkable fact. The Eastern Sea is more than a round, ringed basin; it is instead what geologists now call a "multi-ringed basin." Rather than being a giant crater with a single rim, it has three concentric rims, nestled inside each other.

At the center of the Eastern Sea is a dark, flat plain of frozen basalt about 150 miles across. Surrounding it is a ring of low hills and ridges, interrupted in places by cracks and some lava channels. Beyond this inner ring is a narrow enclosing valley. The eastern portion of this valley is a flat-floored area with a romantic but inappropriate name, the "Lake of Spring" (Lacus Veris). A second ring of hills and mountains surrounds this ringed valley, and together these two circular mountain chains are named the "Rook Mountains" (not named after the European bird, but after the seventeenth-century English astronomer Lawrence Rooke). Beyond the Rook range is another circular valley, wider than the first, with an outer diameter of about 500 miles. A flat portion of it lying over the inner hills from the Lake of Spring is called the Lacus Autumni, the "Lake of Autumn." Elsewhere in this wide valley are several impact craters having the names of famous twentieth-century astronomers, such as Lowell, Pettit, Nicholson, Lallemand and Couder.

The far outer rim of Mare Orientale is a wide, hilly range called Montes Cordillera, which translated into English is "Mountain Range Mountains."

As redundant as Pico Peak, this name seems to "protest too much," as the so-called range consists mostly of a wide system of low, radial hills, largely formed by the splatter of rocks from the great impact itself. The hills and valleys reach out for 300 miles or more from the inner edge of the ring, making the round scar of the Orientale impact have a full diameter of some 1,200 miles.

How is this multi-ringed basin on the other side of the Moon related to the puzzle of Pico Peak? We can probably answer that, but first we must ask whether the remarkable shape of this feature is unique or there are other impact basins with more than one "rim." The crater Schrodinger, located on the lunar far side near the south pole, is an another beautiful example. It has a well-formed outer rim about 190 miles across and nicely nestled in its floor is a low ring of hills about 80 miles in diameter. Schrodinger is an unusually big crater, almost large enough to be called a mare.

Although multi-ringed basins are highly eroded and difficult to see, we also have a few examples here on Earth. The famous Chicxulub crater in Mexico, part of which is under the Caribbean, is about the same size as Schrodinger and it also seems to have a double-ring structure. It is the scar of the impact that is thought by many to have caused giant "tidal waves" that devastated coastal areas and that spread dirt, dust and possibly poisonous gasses over the Earth, killing off many species of life, including the dinosaurs.

Another example on Earth is the Vredefort structure in South Africa. Much older than Chicxulub, Vredefort is also heavily eroded. About half of several concentric rings of outcrops are present. The inner structure can be traced as an arc with a diameter of some 30 miles, with the city of Parys nestled near its center. An outer ring seems to be present at a distance of about 40 miles from the center, according to at least one South African geologist.

Making rings

Multi-ringed basins on Earth are rare only because (luckily for us) large impacts are rare. But geologists find evidence that this kind of ringed structure often results from big impacting events. When the projectile hits the ground a tremendous amount of energy is released. The explosion digs a large crater, throws immense amounts of material into the air and along the surface, shatters and melts the rocks below and sets up tectonic waves that radiate outwards.

No one has seen a multi-ringed basin form on Earth or on the Moon, but the process is not a complete mystery. The various examples seen give up several clues about how the rings are made. Still, there are many questions about the details. Most geologists agree that the outer ring is not a "rim" in the sense of the steep-sized crater walls with overturned rim material that are found for smaller impacts. Instead the basin itself probably was much smaller than the outer ring or rings, which are likely to have been formed by a complex series of events that followed the impact. Perhaps they are "frozen" tsunamis, waves in the crust, like waves in a pond formed by a dropped rock. As rock is stiffer than water, the waves may have frozen in position as they advanced. Evidence also suggests that slumping must have occurred. Perhaps the rings are the tops of giant concentric blocks of rock that were carved by faults and that remained above the rest as the surrounding material slid down into cracks and voids. It is also clear in some cases that volcanic mountains formed along the rings, though the ring mountains themselves are probably not chains of volcanoes. Finally, especially in the cases of the large maria such as the Sea of Rains, flooding clearly occurred, covering up most of the interiors of the basins and possibly burying almost the entire inner ring systems.

Each of these ideas may be important and there also may be other ways of understanding the rings. It is difficult to unravel their mystery from afar. That is why this chapter describes an expedition to Pico Peak, which is probably a fragment of the top of a buried inner ring. By first-hand investigation, accomplished during a mountain climb dreamed of for centuries, you may be able to find the clues that settle the question.

To Pico Peak

The expedition plan assumes that you will be leaving from a lunar base station located near the center of Mare Imbrium. If you and your crew have already been to Mt. Blanc, described in the previous chapter, you will be covering familiar territory as you begin. But as you head out across the frozen basaltic sea, you will be pointing in a more northerly direction. Pico Peak is to the northeast, but your path is not direct, as there are two related features to examine along the way. After a very long day's drive, your first "night" will be spent at a remarkable group of rugged hills

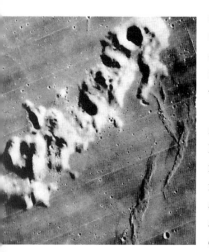

The Straight Range (Montes Recti), a stop along the way to Pico Peak (NASA).

called the "Straight Range" (Montes Recti). Of course, you don't arrive there at lunar nightfall, as the whole expedition must be planned to occur during the 14-Earth-day-long lunar daytime. Your stop at the Straight Range is a sleep stop, mandated by the human internal clock, which remains tied to the 24-hour period of the Earth's rotation.

You also stop here to explore, of course, and the Straight Range is a splendid example of what we think may be part of the buried inner Imbrium ring, of which Pico is such a famous part. The range is about 50 miles long and consists of some twelve or thirteen high hills arranged roughly in a line with a 5-mile crater at one end. The slopes seem to rise abruptly from the surrounding lava plains. From Earth the feature appears to be similar to the Spitzbergen Mountains, which were explored on the way to Mt. Blanc (Chapter 6).

Your expedition should spend a good day examining the range and checking it for evidence about its origin. Are the mountains volcanoes that formed along a crack in the mare floor? Probably not, as their form is too blocky and they appear to be older than the surrounding lava plains. Are they fragments of ejecta from the Imbrium impact that landed here inside the outer ring? Possibly, and a close examination of their structure might support such a view. Or are they the tops of an inner ring, formed by wave action and slumping, later isolated by the flooding of the Imbrium basin? This is what most terrestrial geologists believe and your brief examination may provide on-site evidence for such a conclusion.

After a night's sleep, your expedition will move on to another group of mountains, half-way between the Straight Range and your goal, Pico Peak. These mountains are quite similar to those of the Straight Range, though they reach to higher elevations, topping off at near 8,000 feet. They are called the Teneriffe Mountains and there is an interesting story as to why this name is so appropriate.

The Canary Islands make up a semi-tropical archipelago lying to the west of northern Africa in the Atlantic Ocean. They are volcanic islands, with both ancient and young volcanoes, including one that is frequently active, the high peak called Teide on the island of Tenerife (as it is now spelled on Earth). The islands are part of Spain but they have long been a favorite winter destination for people living in northern Europe and the British Isles. One such person was the prominent nineteenth-century Scottish astronomer, Charles Piazzi Smyth. This unusually forward-looking scientist visited the Canaries with more than a sunny holiday in

mind. He realized that the future of astronomy lay not in the observatories built in Edinburgh, Paris or Berlin, but in the high mountain tops of desert peaks, with their dark nights and clear skies. He carried telescopes to the tops of mountains on Tenerife to test this idea and to demonstrate the spectacular advantages of such an observatory site.

The volcano Teide on the terrestrial island of Tenerife, one of the Canary Islands (author photo).

It took nearly a hundred years for European astronomers to catch up with Piazzi Smyth. Now, however, the mountains of the Canaries have an array of marvelous observatories that enjoy some of the best weather and atmospheric steadiness on Earth. Telescopes built by many countries, including Spain, Great Britain, Germany, the Netherlands and the Scandinavian countries, are clustered on the heights of La Palma, at the edge of a huge volcanic caldera. Another observatory is on the high slopes of Tenerife, in the shadow of steaming Teide. It is thus fitting that the lunar landscape has several features named after these astronomically important islands. In fact, the area of your expedition has four such features. Besides the Teneriffe Mountains, there is Mt. Piton to the east (named after a peak on the island), the crater Piazzi Smyth nearby, and Pico Peak itself, which is named after the highest mountain on Tenerife (called "Pico von Teneriffe") by the astronomer Johann Schroter, who named many of these lunar features.

After an examination of the Teneriffe Mountains and another night's rest, it will be time for your final leg of the trip, the 50-mile journey towards the southeast to the base of Pico Peak. At last you can stand at the base of what has for centuries been the basis for the idea that the lunar landscape is marked by spectacularly sharp spires, rising almost vertically from their surroundings like barren Matterhorns, defying even the thought of climbing them to their dizzying summits. But you will be disappointed.

The climb

Pico Peak stands 8,000 feet above the smooth floor of the Sea of Rains. It is an isolated peak that rises abruptly above the lava fields. The early telescopic drawings of it were made at lunar sunset and sunrise, when its shadow stretched spectacularly across the flatlands, giving the impres-

Near the edge of Mare Imbrium, the Teneriffe Mountains on the Moon were named after the terrestrial island. The lonely peak Pico is nearby (NASA).

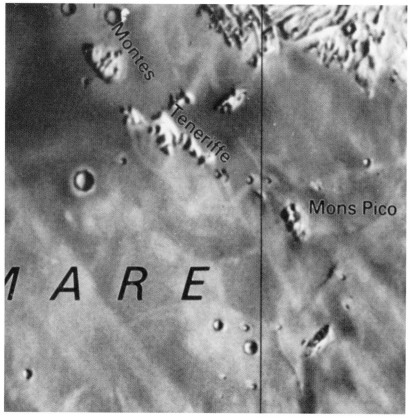

sion of a tall spire. However, when its slopes are carefully measured, Pico becomes a much tamer peak. Its slopes reach angles not of 90°, but about 30°. Your route from its base at the northeast to its summit involves a distance of about 8 miles, so climbing it means gaining about 1,000 feet of altitude for every mile traversed, a quite comfortable ratio, typical of mountain trails in the Alps or the Sierras.

Of course, there is no trail, so the climb will not be trivial. In fact, we know very little about the surface of the peak and there may be unexpected difficulties caused by rough or blocky terrain. But the mountains seen by the Apollo 15 astronauts in the lunar Apennines looked quite smooth and Pico Peak may be similarly uncomplicated. If you have already climbed Mt. Blanc (Chapter 6), you will have first-hand experience with lunar mountains and will know better what to expect on Pico Peak. For planning, you can count on taking a long, full day for the climb, but it may be necessary to break it up into two days. A quick reconnaisance at the base should help you decide.

The route up the mountain will stay at first along the less steep valleys that lead into the heart of the peak from the northeast. When the summit ridge is reached, you will want to bear right to the highest point, which appears from Earth to be a fairly smooth rounded mountaintop just north of the middle of the massif. The view from there should be spectacular.

An imaginary view of the modestly steep Pico Peak above the flat plains of the Sea of Rains.

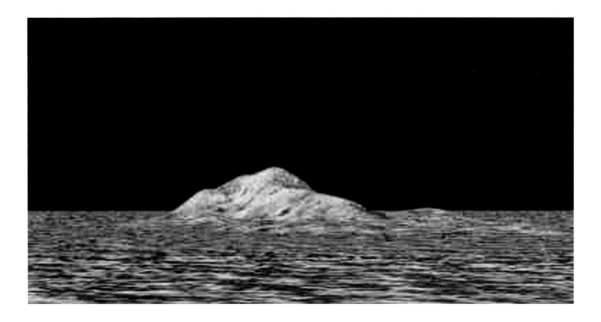

Smooth, flat plains extending all the way to the horizon are decorated by a few small craters and some low hills (called "wrinkle ridges" by astronomers). Even though there are mountains only 50 miles away both to the northwest (the Teneriffe Mountains) and the south (Beta Peak), they are too far to be seen fully, as only their tops peek above the horizon of the Moon.

The scientific purpose of this climb is to gather evidence that will help clinch the theories of the origin of Pico Peak and its nearby siblings. If the mountain is a volcano, you will surely have discovered this in the first steps of your climb, which would in that case be over neatly layered lava flows. The summit should have vents or a volcanic crater. However, no sign of either flows or source has been detected from Earth.

You can also check the suggestion that the peak is a huge chunk of ejecta, thrown out by the impact from a more central location. In that case, you would be traversing a jumble of lunar breccias (mixtures of rocks cemented by impact melt) and blocks of rocks arranged as a helter-skelter pile of rubble.

The more likely explanation of the peak is that it is part of the top of an inner ring, formed by wave action after the impact and by subsequent slumping and inundation. In that case, you should look for remnants of the old, original lunar surface and signs of the faults that formed the ring's shape.

Whatever you discover on the slopes of Pico Peak, your samples and photographs will be eagerly awaited back at your vehicle as well as at home on Earth, where geologists have debated the question of its origin for so long. And you will have conquered one of the fearsome unclimbed peaks of the Solar System, an accomplishment that should not be dismissed just because it turned out not to be as steep or formidable as men and women once dreamed it to be.

8

The great Copernicus traverse

One of the important names in astronomy is that of the Polish astronomer, Nicholas Copernicus. His revolutionary book, *De Revolutionibus Orbium Celestium*, published in 1543, put forward the idea that the planets, including the Earth, all revolve around the Sun. At the time most people assumed in a typically self-centered way that everything – planets, the Sun, the stars and the whole universe revolved around the Earth. Copernicus' idea and his carefully crafted geometrical arguments, though not immediately accepted as truth, eventually led to an understanding of the true nature of the Solar System, and Copernicus was firmly established as one of the founders of modern science.

It is fitting that one of the most glorious examples of a lunar crater was named after Copernicus. It has long been the tradition to name craters on the moon after great scientists of the past. Among the most conspicuous craters we find, besides Copernicus, the names of Archimedes, Plato, Eratosthenes, Kepler and Tycho. Many hundreds of craters have names and it is difficult to think of a famous astronomer of the past for whom there has not been a crater named. Men and women from modern Europe, classical Greece and Rome, the Arab countries, the Americas and the Orient are commemorated in this way. Lunar mapmakers have even slipped in a few non-astronomers, such as the famous chemist Marie Curie and the physicist Albert Einstein. But one tradition has remained almost unbroken: to have a crater named for you, you have to be dead. You can't buy or politically wangle your name onto the Moon.

A very few exceptions to this rule were made in the late twentieth century when a few craters were named after living astronauts, including the lunar pioneers, Apollo 11's Aldrin, Armstrong, and Collins. Their craters lie in Mare Tranquillitatis, the Sea of Tranquillity, near Tranquillity Base, the landing place of Apollo 11, the site of the first human footprints on the Moon.

Forming a crater

A lunar crater like Copernicus is a spectacular feature and an expedition down into its depths is an adventure well worth the effort. But there is

The giant crater Copernicus dominates the view of the center of the Moon (Mt. Wilson Observatory).

more to this motive for such a trip; science does not yet know exactly what happens when an impact crater like Copernicus is formed. The trek up onto the rim and down the formidable cliffs to the bottom will be a wonderful boon to science as well as a thrilling first descent.

Copernicus is about 55 miles across from rim to rim, making it on a par with some of the largest lunar and terrestrial craters, though much smaller than the lunar "seas." In size it's not unusual. What makes it so spectacular when seen from Earth through a telescope is its depth and the starkness of its crisp contours. It is a relatively recent large crater, not degraded by erosion or flooded by lava, and the extensive damage to its surroundings looks fresh, as if the impact just occurred last week.

Of course, it was much longer ago than that. With no water or wind and little geological activity on the Moon, erosion is exceedingly slow, and Copernicus, though recent for a lunar feature, is still almost a billion years old. Your expedition will bring back rocks that will tell us exactly how old by using the technique of radioactive dating.

When Copernicus is viewed from Earth near lunar sunrise or sunset, its floor disappears completely and the center becomes a black void. That is because it is a really deep hole. From a typical spot on its rim to an average place down on the floor, the altitude difference is about 14,000 ft. It is as deep as many famous terrestrial mountains are high. And its walls are steep and structurally complex. The descent will surely prove a challenging and remarkable adventure.

The asteroid or comet that hit the Moon, forming Copernicus, must have been about 5 miles across, the size of a large mountain such as Mt. Everest. When it hit, there was a tremendous explosion. Material was thrown out as from a bomb, creating a giant cloud of rocks that spread across that part of the Moon. Many of these made their own craters as they splashed to the ground, some a hundred miles from the place of impact.

At that point an immense hole was formed, with the rock around it forced up to form a broad rim. The explosion produced shock waves in the ground below, which propagated down and away, shattering the lunar bedrock and melting some of it. That much about the formation we know from the study of close-up photos of lunar craters and from mathematical models of cratering. But that much is not really very much. Many questions remain, especially about the various curious features that close-up photos reveal.

Rock lakes are found both outside and inside the rims of large craters. This example, on the right, lies just below a crater rim and is nestled in the terraces of the crater wall (NASA).

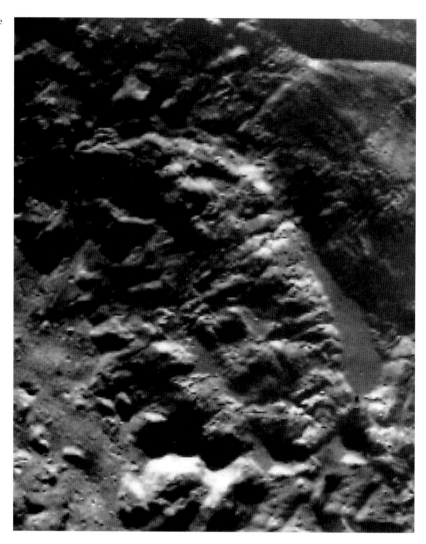

For instance, what formed the crescent-shaped rock lakes that are found near the rim and down along the wall? And what explains the giant terraces of the wall? Why is half of the floor smooth and half hummocky? What is the cause of the range of central mountains found on the crater floor? These and many other questions will be in the back of your mind as you begin the perilous journey to the depths of Copernicus.

The last lunar era

In the early 1960s, pioneer explorers of the geology of the Moon found that they could figure out quite a bit about the history of the Moon using telescopic observations from the Earth. Surprisingly, they found that the Moon gave up many of its secrets when its surface was looked at carefully. The history of the Moon is defined by certain events that affected much of its surface. For example, the mare basins were formed early in time, but their ejected materials can be found to overlie even earlier surface features, the highly cratered and degraded lunar highlands. The mare basins were later filled with lava, which solidified. Later yet more recent collisions made craters in the hardened mare lava. By seeing what lies on top of what, geologists are able to do the ordering in time of the forma-tion of different rock units.

Fortunately, the Moon has a much simpler life history than the Earth has. There is no continent building or plate movements, no oceans, no erosion by water and no wind, so rocks stay where they are, unchanged, until hit by something or flooded by lava.

Eugene Shoemaker and his colleagues at the US Geological Survey were among the first to divide up lunar history into its different periods. They worked before the Apollo and Luna programs brought back lunar rocks for age-dating, so their dates were relative dates, much as the Earth's were before isotopic analysis made age estimates more precise. But their results were clear and have held up well over the years.

The earliest era in Shoemaker's "stratigraphic column" (an imaginary column of rock in which each rock-forming event is identified, from the bottom, the oldest, to the top, the youngest) is called the Pre-Imbrium Era. It was the time when the cratered highlands were formed and when the older mare basins were created, the Imbrium Basin being younger than the others. In the years since the 1960s lunar geologists have further refined the sequence of events of this period, but those details primarily affect other parts of the Moon, far from Copernicus, our topic in this chapter.

The second era in Shoemaker's scheme is called the Imbrium Period. It is defined by the creation of the huge Imbrium basin and its subsequent flooding by lava. The Oceanus Procellarum (Ocean of Storms), a vast mare surface west of Mare Imbrium, was also flooded at about this time.

The third period is called the Eratosthenian, named after the large crater Eratosthenes, which lies on the southern "shores" of Mare Imbrium, northeast of Copernicus. Eratosthenes and the many other craters of this period are sufficiently old that they no longer have prominent rays of ejected material or secondary craters (produced by chunks of rock ejected by the impact), all of which have been worn away by the erosion of subsequent smaller impacts. But the Eratosthenian craters lie on top of the mare lava flows, so they are younger than the Imbrium Period. Other spectacular craters of this period are Langrenus, Herschel and Aristoteles.

Finally came the Copernican Period, when rayed craters like Copernicus were formed. These are relatively young and uneroded, with extensive displays of secondary craters and rays, some of which extend out across the surface for hundreds of miles. Other Copernican system craters are Kepler to the west and Tycho to the south, both named after astronomers who lived and worked just a little later than the time of Copernicus.

The rocks of Copernicus

What will you look for and find as you travel up to the rim of Copernicus and then down into its vast crater? At the time of writing, no one has set foot on Copernicus and your expedition may be the first to visit it. Most of our information about its rocks comes, therefore, from telescopic and lunar orbital examination. However, the Apollo 12 astronauts landed about 250 miles south of Copernicus, on (or at least near) one of its rays, and some of the rocks they brought back may be ejecta from that crater.

The currently best age estimate for Copernicus comes from material dug up by those astronauts from the area of the ray. This material was age-dated using the argon gas that it contains, part of which is naturally radioactive, with a resulting age of about 800 million years. Although this sounds old to our ears, it is really quite young for the Moon, most of which is covered with rocks that are more than 3 billion years old.

Because the Moon has a rather simple history, there are only a few types of rocks on its surface. The rocks that you will encounter will probably be of only three types. Most common will be dark basalt, volcanic rock that formed when lava filled the lunar basins. This basalt is very similar to Earth's basalt, which abounds in places like Hawaii, India, central Oregon, and hundreds of other volcanic regions. The rocks are black, and sometimes cindery and sharp-edged.

A microscopic view of a thin section of lunar basalt, showing the various sharply shaped mineral crystals.

Some lunar rocks have an unusual composition, not found on Earth; these are the KREEP samples, so-called because they are abnormally rich in potassium (chemical symbol: K), Rare Earth Elements (REE) and phosphorous (P). They have unusually-high percentages of barium, rubidium, zirconium and other similar elements, though these are still only very minor constituents. The supposed Copernican samples dug up by the Apollo 12 astronauts were rich in KREEP material.

Did the lunar basalts issue forth from cracks in the crust or were they formed by the melting of the surface rocks caused by the impact of meteorites? Either process involves molten rock that spreads over the surface and cools. Geologists believe that both kinds of basalt are present on the Moon, but that impact melt probably dominates, as suggested by trace amounts of meteoritic material that seems to contaminate much of the lunar basalt.

Another common lunar rock type is impact breccia. This will be quite abundant around Copernicus. Breccia is formed from fragments of various rocks that have been cemented together by glassy material

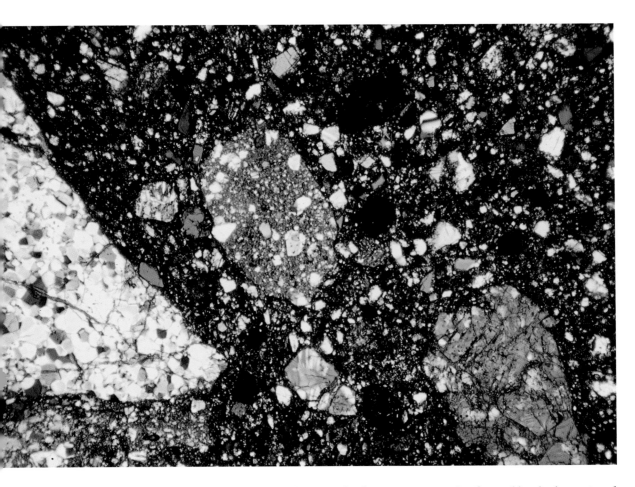

A microscopic view of a thin section of lunar breccia, showing small pieces of various kinds of rocks all captured in a matrix of impact melt rock.

melted by the impact. The fragments were also formed by the impact and range in size from tiny grains to large, hand-sized chunks. Breccia is generally lighter in color than lunar basalt but, to identify it for sure, it is best to examine thin sections of it under a microscope, which show nicely preserved irregular chunks of rock, all captured in a smooth, glassy matrix.

The rarest rock type, unlikely to be found in abundance among your Copernicus area samples, is the ANT series, which include Anorthosite, Norite and Troctolite, three rock types of an igneous nature also found on Earth. The ANT rocks are the oldest lunar samples and probably date back to the era when the lunar surface was largely or wholly molten. The geologists on your team will find these names to be quite familiar, but the rest of you may prefer not to get too involved in the complicated terminology of lunar petrology (the science of rock types). For example, a description of one ANT sample from near Copernicus goes like this: "a pristine pink spinel troctolite clast with euhedral to subhedral olivine enclosed poikilitically by plagioclase."

The traverse

Copernicus lies near the center of the lunar Earthside in an area rich in craters of various ages. To its east is the "ghost crater" Stadius, almost completely inundated by lava with only the walls of its rim peeking above the volcanic plain. Farther to the north is the spectacular crater Eratosthenes, almost as large as Copernicus, but older, having given its name to the geological period that occurred just before the Copernican. South of Copernicus is another nice crater of the Eratosthenian Period named Rheinhold.

The plan is to drive south from the lunar base in Mare Imbrium to the smooth floor of mare material that lies between Eratosthenes and Copernicus. This journey can occupy the first day of the traverse, with the night (a terrestrial night, of course, as it will be daylight here on the Moon throughout the trip) spent at a nice flat area at longitude 15°W, latitude 14°N.

The second day's drive will be a slow one, as you will make frequent stops to examine and gather samples of Copernicus' famous system of rays and secondary craters. You will be starting 75 miles from the rim of the crater.

This day begins with a drive to the southwest through a field of small craters and Copernican ray material. The craters include many secondary craters, formed by chunks of lunar material thrown out by the great impact that formed Copernicus. You will want to examine them, as they are unique. No secondary craters have been preserved around Earth's meteorite craters and so we have not had much of a chance to study this phenomenon up close. They should look quite different from similarly-sized primary craters, as the projectile that made them came in at a low angle and at low velocity. The mapped craters here range from 2 or 3 miles in diameter down to less than half a mile, but you will no doubt find hundreds of smaller craters, too small to show up on today's maps. Some of these may be shaped like long narrow trenches and they may still have the rock in them that formed them. Sampling and careful mapping will be important tasks on this second day.

The rays themselves look from Earth and from lunar orbit like a thin covering of bright splattered material, probably including fresh crushed rock from the impact. We don't really have good samples of it, however, and so don't know how big the fragments are. Do they consist mostly of sand-sized grains or are there lots of big chunks, perhaps the size of a house? You will be the first to explore these rays and to answer the question definitively.

But there are other, more mysterious features around you. The day begins in the "third zone" of the flanks of the crater, the outermost zone of the ejecta blanket. Perhaps the weirdest features of this zone are the "herringbone" craters, spread over the surface in a regular design with the tips of the angles pointing away from Copernicus. The "backbone" of the herring is a nearly-linear groove with a raised rim and the "ribs" are lines of light-colored material radiating away on either side. How did these weird patterns form? No one knows but there are some conjectures. For example, they may be the result of a collision between two sheets of ejecta above the surface or they may be formed somehow from the base surge of material along the ground. Your visit should help resolve this mystery.

You soon reach the "second zone" of the crater flanks. It extends from about 50 miles from the rim up to about 15 miles from it and has several interesting features to explore. There are many narrow hills and valleys that point out away from the crater ("radial lineations" geologists call them) and there are smoothish areas where the land is hummocky. Here and there one can find small flat pools of frozen lava in depressions. Were these pools formed from liquid rock that splattered out from the impact or were they formed from below, where rock was liquefied by the intense pressures and then issued up through cracks to fill small basins? You'll find more of these pools on the way down into the crater.

The second zone is also marked by crescent-shaped ridges of ejected material, arranged with the horns of the crescent pointing away from Copernicus. These were probably sheets of rock fragments that were spewed out and remained together until they splashed down onto the surface.

The "first zone" of the crater surroundings extends from the rim out about 15 miles. As your vehicle bumps along into this area you will notice an increasingly rugged terrain and many huge boulders. Deep valleys and ridges, much larger than found in the outer zones, radiate away from the crater. Some are as much as 5 miles wide and some turn into huge concentric furrows running parallel to the rim. The last few miles to the rim will surely tax your vehicle and its driver as the route takes you up and down through a series of giant corrugated annular fractures. But the average trend will be *up*. The crest of the crater rim at its highest point is nearly 3,000 ft above the level of the surrounding mare surface.

On the rim

The plan is to spend the second night on a high knoll on the northeast rim of the crater (latitude 11° 04' N, longitude 19° 14' W). From this vantage point it is possible to see out over the plains to the north and east and down into the crater to the southwest. A day should be spent at this locale to allow time to explore the many rim features and to work out the best route for the steep descent.

The rim of Copernicus appears to be a very rugged place. The high points have sharp, angular material, as if any fine debris from the impact was swept clear by the blast wave. Many faults are visible, arc-shaped and concentric with the crater walls. These cracks in the surface show evidence of slumping, probably the result of the relaxation of the rock after the violent upheaval caused by the collision. Most of what you will see around you has remained much the same over the hundreds of millions of years since the crater formed, but not everything. Perched on the rim and visible from your knoll about 3 miles to the west will be a small crater, nearly a mile in diameter, that resulted from a more recent meteorite impact.

Down the terrace walls

The main challenge of this expedition begins on the fourth day. Having worked out a promising route down into the crater the previous day, the climbers will be ready to descend the steep cliffs of the inner wall, from terrace to terrace. The distance "as the crow flies" to the bottom is only 15 miles, but the vertical distance from this point on the rim is about 17,000 ft, a staggering height considering the steepness of the walls. To have both a safe and a scientifically productive trip, you will want to spend a good three days in a descent, with another three days to cross the crater floor to the central mountains and four to six more days for the return to the rim. Provisions for all but the first three days will have to have been sent to the crater floor by a soft-landing rocket in advance. You'll need supplies of oxygen, food and water.

The first day will probably be the most challenging. The walls of the crater are in the form of huge steps. Steep scarps alternate with nearly flat terraces, each step the result of massive faulting of the lunar rock and subsequent slumping. Dotted here and there on the terraces are smooth, flat pools of once-melted rock. These pools were probably formed by drainage of liquid rock from the slopes above or by upwelling of molten material from below. Close examination right there will probably settle that question.

The steep cliffs that make the top of the crater walls so daunting have a few places where a reasonably safe descent should be possible. Your probable route will take you down to the southwest about 5 miles to the brink of the uppermost cliff. A steep valley leads down from there to the southeast and might provide a good, though still probably treacherous, way to descend. As you go, you'll be climbing down through several interesting layers of rock. The upper layers are probably the edges of the blankets of ejected and overturned material from the impact, sheered off as by a knife when the ground cracked and the inner parts slid downslope. Below that you will probably come to layers of the pre-impact lunar crust in which the earlier history of the Moon is laid out. Exploring these outcrops of rock may make you the first persons actually to see the lunar geologic column. Chapter after chapter of lunar history will be revealed under your feet as you descend into the Moon's distant past.

As in many cliffs in the Earth's mountains, the crater scarps have messy jumbles of rock at their bases. These are the chunks of rock that fell from above after the slumping stopped. Finding a good route through this massive talus may require as much heroic effort as was needed in descending the cliff itself.

The first night on the wall is probably best spent on the welcoming flat floor of a frozen pool. You may choose the one directly south of the peak where you left the lunar vehicle. It is about 10 miles by foot from the starting place and is not only pleasantly flat, but is also a goal in itself, as you can spend the "evening" after dinner exploring its shores, looking for clues about whether the liquid lava drained down from above or issued out from below. This is a nice big lake, over a mile across.

The next day will take you down through more rugged terrain and along a radial trough that cuts through the series of cliffs. This narrow valley may be a fracture or it may be a channel formed by the flow of liquid or partially fluid material. The second night can be spent at another nice pond. This one is 11,000 feet below your starting point high on the rim.

The third day brings you at last to the crater floor. The trip follows a wide channel, skirts along the eastern shore of another rock pond, passes a curious round hill on the right, and finally leads out onto the flat crater floor.

Across the floor

The floor of Copernicus is not a simple place. Instead of a nice, smooth flat plain, as found, for instance, in smaller craters like the one in Arizona, Copernicus' floor is amazingly varied in texture. There are areas where fractures form a maze of narrow valleys and ridges. There are mounds that resemble volcanic cones, some with craters on top and some with moats or radiating fractures around them. Some of the circular features are rings of material, looking very much like "tuff rings," found in some volcanic areas of the Earth and made from eruptions of frothy, viscous material that falls to the ground as light granular deposits. All of these various features are spread out over the huge area of the floor, which occupies about a thousand square miles.

Following another night's rest period, it will be time to set out across the crater floor. Your ultimate goal is to reach the central mountain peaks, which are a marvelous geological treasure. Hidden in those mountains are many secrets — secrets about how giant craters form and secrets about the early Moon.

But first you must get there and they are a long way away. From the bottom of the crater wall to the center of the central mountains is a straight-line distance of about 20 miles. You will want to spend two days crossing the rugged and interesting terrain to the foot of the peaks and another day exploring them.

The route begins easily enough on the rolling ground that is typical of the northern half of the crater floor. Smoother than the rest of the plains, this surface is thought to be made up of rock that was melted or partly melted by the impact. As it solidified, more melt flowed down from the heights, creating channels and ridges. The mushy, partly frozen rock

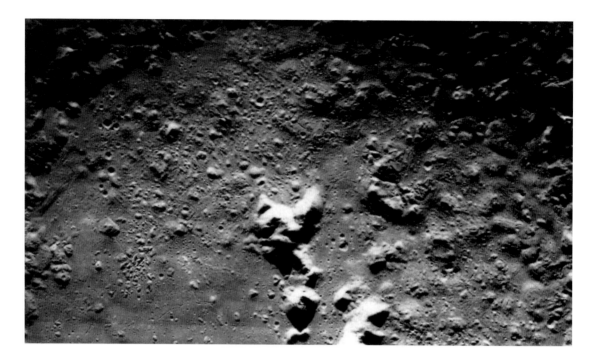

The central peaks and part of the floor of Copernicus (NASA).

cracked here and there, leaving fissures and troughs that you'll have to wind your way through or around. These cracks are tens of feet deep and a hundred feet or so across.

Three miles from the base camp at the foot of the wall is a change of terrain. You'll find yourself entering a field of circular craters. Close-up orbital photos show hundreds of perfect, round craters, spread out as if spattered from above. They can't be just random impact craters from the subsequent impact history of the Moon, as there are too many of them and they are too localized. Geologists think that they may be secondary craters formed by a cluster of material that was ejected nearly straight up. The chunks of rock must have almost left the Moon into space, their velocity being close to the escape velocity, which is only about $1\frac{1}{2}$ miles per second on the Moon (we have found several meteorites on Earth that are rocks ejected like this by impacts from the lunar surface). Then they fell back to the crater, which had by that time, several minutes after the impact, settled down to nearly its final shape. The vertically falling rocks made nice circular pits, which you will pass on your journey south.

The second day on the crater floor will find your expedition moving across the second type of terrain, a type that occupies a bit more than half of the floor, mostly in the south. This may be hard going. It consists of rough, hummocky country, pocked by close-packed rounded hills. There are big, blocky hills up to 2 or 3 miles across, as well as smaller ones, all surrounded by bits of the flatter textured material encountered earlier. These hummocks are probably chunks of bedrock, fractured and displaced by the impact and coated with a frosting of impact melt.

As you find your way across the floor you will probably encounter many other remarkable features, maybe some that are unique to this type of formation and that can't be found in geology textbooks. But as fascinating as these things may be, you will probably feel the urge to push on quickly as the ultimate goal of the expedition looms above you.

Actually, "looms" may not be the best word here. The central mountains are indeed high. The highest is nearly 7,000 ft higher than the lowest place on the crater floor. But the profiles of the peaks are probably not very steep. They are more like the White Mountains of New Hampshire than the Alps of Switzerland. Nevertheless they represent a fine case of a remarkable "first ascent" for the climbers in your party. Never before will men or women have summited a mountain like this, buried as it is in the bottom of a vast hole.

The mystery peaks

The central peaks of lunar craters remain a source of mystery and intrigue. Most scientists believe that they are huge chunks of rock that rebounded from below after the impact, like a bounce. But careful examination of Copernicus from orbit suggests that there may be another explanation. Not all of their features discernable so far are easily explained.

There are three main theories put forward to explain these peaks. First is the idea that they are simply large masses of rock that rebounded. A second theory is that they are the toes of massive slumping, in which immense avalanches of material flowed down the walls of the crater to meet in the middle, leaving a jumble of debris there. A third possibility is that they are volcanoes, formed from subterranean melt produced by the force of the impact.

The first theory is generally favored and it does seem to have some good arguments on its side. For example, computer simulations and laboratory experiments suggest that a rebound should occur, involving a large upward displacement of material. There are outcrops of rock on the sides of the mountains and blocks at their feet that appear split into slabs, as if they are made of the ancient layers of lunar basement rock. Furthermore, geological exploration of large terrestrial craters, such as the Sierra Madera structure in Texas, also supports the idea.

The theory that they are the toe of concentric avalanches seems preposterous, perhaps, but the central location and the irregular and rugged shape of the peaks might argue in its favor.

On the other hand, there are also clear suggestions that the mountains may be volcanoes. Lunar Orbiter close-up images show summit craters on some of the peaks and there are furrows on the mountain flanks that resemble volcanic fissures. Between the peaks there are small craters with smooth, subdued surfaces, looking as if they are covered with something like volcanic ash.

Well, which is it? Surely your party of explorers will find a wealth of evidence that should solve this mystery. Even only a day or two on the slopes of these remarkable peaks should provide an exciting and revealing experience, an example of a unique first ascent.

Maxwell, mountains of mystery

Venus comes closer to Earth than any other planet, but in spite of that its surface has remained hidden forever from terrestrial eyes. Perpetual clouds cover the planet so completely that there is never a clear day — never a parting of the veil to let us see what's down there. After searching Venus with telescopes for nearly four centuries the eyes of Earth have never glimpsed the face of its twin.

And Venus *is* Earth's twin. Its size is nearly the same and its mass and density are similar to Earth's. But it is certainly not an identical twin. It presents a terribly different environment, one so harsh that it will be a courageous party indeed who first tries on foot to explore its surface.

Like Earth, Venus has an atmosphere. Unlike Mars, whose atmosphere is very thin, the atmosphere of Venus is much thicker that Earth's, nearly a hundred times thicker. It consists almost entirely of carbon dioxide, with virtually no oxygen, so it cannot be breathed. What's worse, it is incredibly hot near the surface, where daytime temperatures peak at about 900 °F. Furthermore, there is corrosive acid in the upper atmosphere; some of the clouds are made up at least partly of sulfuric and hydrochloric acid. Explorers to the surface of Venus will need a very elaborate protective suit to withstand the horrible conditions there. All of the successful spacecraft that have landed on the surface so far died within an hour or so, in spite of being designed with great care. Living on Venus for even a short time will be a tremendous challenge.

And yet people will go to Venus. It is so close and its surface is so exotic that we cannot possibly resist the temptation to explore it. You might as well be among the first.

The surface revealed

Although no people have seen the surface of Venus with their own eyes, spacecraft have now mapped it fairly well. In the early 1990s the Magellan Orbiter used radar to produce a beautifully detailed map of the entire surface. A wild and wonderful place it has turned out to be, harsher and more foreign than anyone expected. It is covered with features, large and small, many of which are unique to the planet.

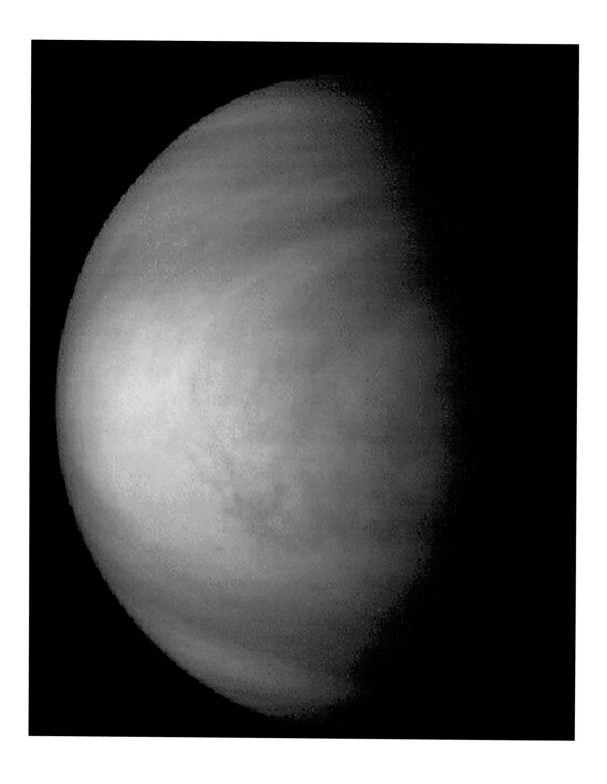

Among the weirdest are the strange things that seem to be exotic volcanoes. Some are giant pancake-shaped hills, each of which may have been formed by a single volcanic burp. Others are huge elliptical formations, called coronae, which seem to be giant failed volcanoes, where the lava pushed up, distorting the surface, but never actually made it out onto the land to make a proper volcano.

There are many other strange features on Venus, some of them probably related to the stress and strain of the shifting of the crust as it floats above the molten interior. Some areas are cut by hundreds of linear faults, making a fearsome landscape called "graph-paper terrain." Related to this are the huge areas of jumbled ridges called "tesserae," where travel would be incredibly difficult. The same shifting forces apparently have pushed up the crust to form several continent-sized highlands, the highest of which are higher than Everest. It is the highest of these barren and rugged peaks, the 40,000 ft Mt. Maxwell, that is the goal of this chapter's expedition.

I would be badly misleading you if I told you that I am firmly convinced that humans will soon walk the barren landscape of Venus. It is not easy to imagine how a safe "Venus-suit" might be built. But it is also true that human ingenuity is pretty amazing. I remember that my grandfather was brought up in a world without automobiles and, although he eventually bought an early Ford and actually tried to drive it, he never admitted that airplanes didn't disobey the laws of physics, refusing to even consider riding on one. And for him space travel was completely impossible, it didn't even deserve thinking about.

Because of my (perhaps unnecessary) doubts, this chapter chooses not to prescribe a land-based expedition to the summit of Mt. Maxwell, involving a heroic and maybe impossible trek in Hell. Instead, it chooses a safer and more comfortable way to the summit, descending in a Venus-style "submarine" (but see Chapter 20).

The land of the goddess of love

The Maxwell Mountains rise above a magnificent, high plateau on one side and a jumbled land of saw-toothed ridges (tesserae) on the other. The whole area, including other, lesser mountains, is named Ishtar after the Babylonian goddess of love. These high lands were among the first places to be detected on the surface of Venus, having been seen as a bright spot in

[opposite] Venus seen from above its atmosphere shows only the top layer of thick clouds (NASA).

Unusual features on Venus' surface include these volcanic domes called "pancake hills" (NASA).

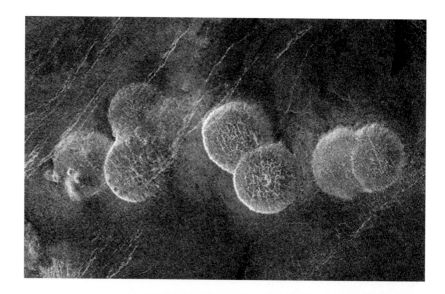

Venus has many large volcanic features called "coronae." The black lines are data artifacts (NASA).

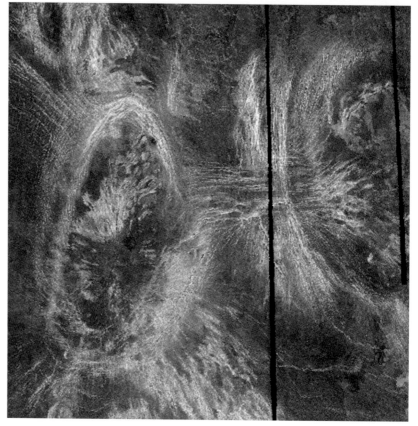

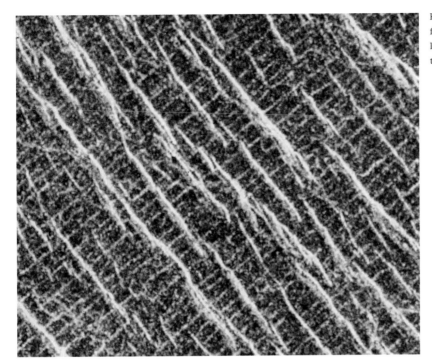

Remarkably cut by crossed faulting of the surface, this Venus landscape is called "graph-paper" terrain (NASA).

the reflected radar signals that bounced back from the planet to Earth. The first good Earth-based radar maps of Venus, made in the 1970s, revealed several of these highly reflecting areas, the brightest being in the high north. Because it was detected using radio waves and because radio waves and other forms of light were first described mathematically by James Clerk Maxwell, a nineteenth-century British physicist, scientists named this bright object "Maxwell." This was before it was found to be a great mountain massif of staggering height, as well as before the International Astronomical Union had decided that all names on Venus should be feminine.

Except for a few other early discoveries of vague landforms ("Alpha" and "Beta," for instance), all of the nearly 1,000 named features on the surface of Venus have feminine names. Some are just first names common in various cultures — for example, Almeida (Portugese), Evika (Tartar), and Maa-Ling (Chinese). Many are names of female goddesses related in some way to Venus — for example, Khotun (the Jakut goddess of plenty), Aranyani (an Indian goddess of the forest), and Kara (an Icelandic valkyrie). But most features, especially the craters, are named after famous women of the past, whose accomplishments range from science to politics to the arts — for instance, Sikibu Izumi (a tenth-century Japanese writer), Carmen Amaya (a Spanish Gypsy dancer) and Susan B. Anthony (an American suffrage leader).

Ishtar is an immense and complicated land, perhaps the closest thing that Venus has to being a true continent. With its steep mountains and high plateaus, it rises far above the surrounding plains. In fact, it rises much too far above them. Scientists are mystified by Ishtar; their calculations indicate that it shouldn't exist. If its heights were pushed up by crustal crowding or plumes of molten rock from below, then it should have sunk again to much lower levels because of its great mass. The plastic material below Venus' torrid surface shouldn't be able to hold such high mountains up. Is it still being pushed up? Are the high peaks and complex clusters of huge cracks recent results of some mutant form of continent building going on before our radar eyes? We don't know now and may have to wait until someone (like you) or some thing goes there, examines the ground in place and brings back samples to study.

Fortuna the unfortunate

The eastern side of Ishtar is one horrible mess. Extending for 4,000 miles from the Maxwell Mountains is a jumbled high plateau called Fortuna Tessera, a land of ancient wrinkles that may be among the oldest surfaces on the planet. These lands, as large as Europe, appear to have been squashed and stretched and split by eons of crustal movement. The rugged and desolate country is made of hundreds of ridges and valleys and several huge canyons, hundreds of miles long. Perhaps it is a rare example of the old Venus surface that everywhere else seems to have been covered by lava from a long-ago event. Its age may be billions of years, its wrinkles the result of the strain of eons of pushing and pulling. Or maybe not. Tesserae are a fascinating mystery and the large example in eastern Ishtar, named as it is after the Roman goddess of chance, appears battered by misfortunes of a long and violent history.

High plains of the war goddess

Mt. Maxwell falls off surprisingly steeply on its western side. In only about 100 miles from the highest summit to its base, the altitude drop is nearly

20,000 ft. There are sections of this west face that average a steep 35° or more. Of course Earth mountains have individual cliffs much steeper than this. The famous North Wall of the Eiger in Switzerland, for example, is nearly vertical. But it is only 5,000 feet from top to bottom, only a fourth the height of Maxwell's western face, and only a few hundred feet wide.

The foot of Maxwell is not at Venus' version of "sea level." As in the case of the Eiger, the mountain rises up from high ground. But unlike the Eiger's base, this high ground is not the lush flower- and glacier-dotted valleys of the Bernese Oberland. Instead it is an immense, flat and desolate plateau, larger than Tibet, higher and much hotter.

Named "Lakshmi" after the Indian goddess of love and war, the high plateau is ringed on three sides by high mountain ridges. Maxwell towers above its eastern edge while the Freya and Akna Mountains form a protective wall on the north and west. The southern side of the Lakshmi Plain is a scalloped terrace with a steep drop to the lowlands 10,000 feet below. Directly beneath these cliffs the Venusian plains are rough and hilly as if landslides had carried rocks from the slowly rising heights out onto the surroundings. The straight part of this steep wall is called Vesta after the Roman goddess of the hearth.

The Lakshmi Plain is smooth and flat. Early radar maps failed to explain why. What strange elevator-like force could raise an immense section of land up so high, keeping its surface from being all wrinkled and hummocky? The details finally sent to Earth by the Magellan spacecraft provided a partial answer. The plains turn out to be a massive field of volcanic flows, fed from several immense volcanic craters. The huge calderas called Sacajawea and Colette, originally thought to be large impact craters, were revealed to be volcanic instead and are probably the source of Lakshmi's vast, smooth fields of frozen lava. Each caldera is about 100 miles in diameter and a few thousand feet deep, huge by the standards of any planet in the Solar System.

Ship of the clouds

The Maxwell Mountains are a formidable goal. Higher than Everest, hotter than Hades, steeper than the Alps, and surrounded by worse air than found in even Earth's most polluted cities, it presents a real challenge. How to reach its peak; how to summit this top of the planet sometimes called, inaccurately, Earth's twin? The easy way is not to ascend its heights, but to descend. Some day people will be exploring Venus in person, but at first, at least, these visits will not be to Venus' horrible surface, but rather to high in its atmosphere. There will be scientific colonies there, housed in comfortable, air-conditioned gondolas hanging from giant balloons. At altitudes of about 150,000 feet above the hot Venus surface the air pressure is reasonably like that at Earth's surface and the temperature is a moderate 120 °F, like that of a summer day in Phoenix, Arizona. Your best bet, if you want to be the first humans to set foot on the desolate Maxwell summit, is to arrange for it to be a careful descent from one of these comfortable exploration airships.

The scientists of the airship will probably be anxious to support your endeavor. It is unlikely that anyone will have attempted to land on the Venus lowlands at the time of your visit; the pressures there are too high and the temperatures too hot. However, conditions on the summit are less extreme. The air pressure is still quite high, of course, but the temperatures will be as low as a "balmy" 700 °F. Of course, this is still quite extreme, but any environmental suit that you might have brought along is likely to be more forgiving under these comparatively less demanding conditions.

The plan will be for your small party to leave the airship in a spherical vessel that is rather like the exploration submarines that have been used for human travel to Earth's ocean's depths. This small craft will need thick, well-insulated walls, portholes for viewing, a door, a propeller, an anchor, and a supporting balloon. Its descent will be slow, as it must travel both sideways and down, sideways because you will need to compensate for the high, steady winds in Venus' upper atmosphere. If the descent is made at the modest speed of about 2 miles per hour, you can expect to reach the summit of Maxwell a mere 12 hours after you leave the balloon-borne colony.

By the time you near the altitude of Maxwell's summit, your crew will be ready for its rest period. It won't be night, though. Venus' slow rotation period, 243 Earth days (backwards), means that your entire expedition can take place during one of Venus' days. Because its rotation period is similar to its orbital period, 225 Earth days, any place on Venus experiences daylight for just over 58 Earth days, followed by an equally long, dark night. But we Terrestrials are biologically set to the clock of Earth's day and night, so the crew had better have a nice dinner and a good rest

before trying the delicate maneuver of landing on Venus' horrible heights.

A snowless summit?

Finally the big day will come. Your crew will carefully guide your vessel to the highest point on the Maxwell Mountains. What will these high desert slopes look like? We don't know. But radar images of the high mountains of Venus look almost as if they are snow-covered. Compared with the lowlands, the peaks of Venus are shiny. Everywhere above about 12,000 ft on Venus there is some snow-like substance that reflects radar waves better than elsewhere, making the maps look almost like visual photos of

An imaginary view of Mt. Maxwell.

the Earth's mountains as seen from the Space Shuttle. But on Venus it certainly can't be snow, of course, as it's far too hot and dry.

We know what it isn't, but we don't know what it is. Possibly before your trip, some sort of remote sensing will have solved the mystery and the scene below you will not be a great surprise. Now, however, we can only speculate about the snows of Venus. Two remarkable ideas have been put forward to attempt to explain them. Astronomers have tried to think of some substance that would condense onto the mountaintops like snow but would remain in vapor form at hotter, lower elevations. The "snow-line" on Venus, regardless of latitude (which makes little difference on Venus) is located where the temperature has dipped to about 820 °F.

Several materials might fit the bill. For instance, lead and copper would be liquid down on the lowlands, but could be solid on the cooler heights. But the substance that seems to be just right is a much more exotic element, the rare earth named after the Earth, tellurium. Some scientists have pointed out that tellurium, if present in the atmosphere, would condense out, forming a weird frost at just the right altitude. But is there enough in the atmosphere to coat all of the mountains of Venus? Maybe, just barely. If the volcanoes on Venus vent out similar gasses to those from terrestrial volcanoes, there might just be enough tellurium to cover the rocky peaks. It would be a very thin layer, only perhaps a micron (0.0004 inches) thick, but calculations suggest that even such a layer might cause the observed high radar reflectivity.

Another suggestion is that the mountains are covered with "fool's gold." This shiny, yellow stuff is made of iron and sulfur and is properly called iron pyrite. Some astronomers have pointed out that at the temperatures above Venus' snowline iron could chemically combine with the sulfur in the atmosphere to cover the peaks with a sparkling layer of this gold-colored mineral. It may not be stable in Venus' atmosphere, however, but some scientists think that it is more likely to be there than the more rare and exotic tellurium. Perhaps Venus is gold-rimmed like some expensive china pattern.

In any case, the snow is probably a thin layer and it may not even be conspicuous as you peer out of the porthole, as the mountaintops may be dust-covered. Radar penetrates below any surface dirt to see whatever lies just below.

Once your vessel has reached the mountaintop, your crew will want to drop anchor and secure the ship to the ground. Once the vessel is settled, the crew can prepare to move out onto the rocks of Venus' highest mountain. Extremely elaborate "space suits" will be needed to protect you from the hotter-than-oven temperature and the time outside the safety of the craft will wisely be made short on this first trip to the surface of such a hostile planet.

There will be time, though, to take lots of pictures and to gather lots of samples to bring back. Will you, like a disappointed old-time prospector, bring home a sack full of fool's gold? Perhaps, or maybe your bag will hold more exotic materials. In any case, the rocks that are brought back up to the "Ship of the Clouds," along with your exhausted crew, will provide scientists with a marvelous opportunity to piece together the geological puzzle of the high peaks of Venus, the Maxwell Mountains, higher than Everest.

10

Volcanoes of Venus

As explained in the last chapter, in some ways Venus is Earth's twin and in other ways it is spectacularly different. One difference is that, lacking oceans, it has three times as much land surface. Thus we should expect more of everything, more mountains, more plains, more valleys, and more variety. Indeed, Venus does have more of lots of these. But the most conspicuous richness found on Venus is its plentitude of volcanoes. Nearly 90% of the surface of Venus is volcanic in nature.

Normal and bizarre

Among the volcanic features on Venus are many normal things, such as lava flows and volcanoes. The lava flows often appear to have issued from cracks in the surface and spread out over the plains, as flows have done on Earth. Some have formed long lava rivers, which are a little more surprising, and a few of these have just kept on flowing, forming frozen

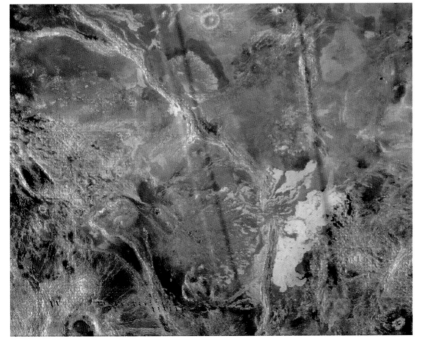

Lava flows on the surface of Venus. Smooth lava (probably like terrestrial pahoehoe) shows up as bright areas and rougher lava (probably like terrestrial aa flows) shows up as dark in this radar image. The slanting parallel lines are data artifacts (NASA).

rock streams that are thousands of miles long, more extensive than even the longest of Earth's water rivers, such as the Nile or the Amazon. This may be at least partly the result of Venus' hot surface, where low-viscosity lava can travel a considerable distance before cooling and hardening.

Many of Venus' thousands of volcanoes look normal to terrestrial eyes. There are many small cinder cones, especially notably grouped together

Arachnoids on Venus (NASA).

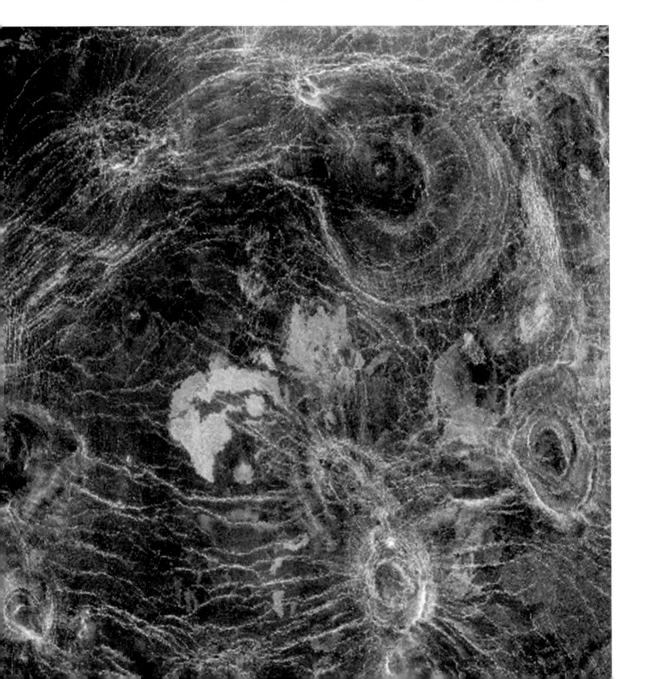

in what are called "shield fields," which are about 100 miles across and which can contain hundreds of small volcanic cones. There are also many larger volcanic mountains, most of which resemble "shield volcanoes" on Earth, gentle-sloped smooth mountains like Mauna Loa in Hawaii or Olympus on Mars, though none is as large as either of these. Most are about the size of Mt. Etna on Sicily. One of the highest of them, Mt. Tepev, will be the ultimate goal of this chapter's planetary adventure.

Venus' rich variety of volcanic features also includes some things that are truly weird. There is nothing on Earth that resembles, for instance, the "arachnoids," strange circular features that vaguely look like giant spiders. They have central mounds tens of miles across, which are ringed by circular depressions or cracks in the surface, from which radiate long linear faults. Other things called "novas" have similar sizes and similar "legs," but no "body." The long lines of faults seem to emanate from their centers as in a photograph of a bright new star (a nova). Geologists who study these strange objects are convinced that they are volcanic in nature, produced in some way by upwelling from below. But there is nothing presently on Earth with which to compare them.

Novas and arachnoids are apparently related to a class of remarkable volcanic features called "coronas." These are giant domes, circular or elliptical in shape, with diameters averaging 150 miles or so. Their centers are only a little higher (sometimes lower) than the surrounding fields and their edges are often marked by concentric cracks and a depression that is something like a moat. The low dome of a corona is usually unbroken by any crater and unmarred by any lava flows. Students of coronas believe that they are failed volcanic mountains, where a blob of molten rock rose from below, pushed up against the surface layers forming a stretched but unbroken mound there, and then stopped and receded, leaving the flattened dome and the surrounding stretch marks.

In the Plains of Niobe

The exploration of the surface of Venus will not, at first, be on foot. This chapter outlines a plan to visit some volcanoes and related features, but first from the air. With a base station in a balloon-borne craft high above the surface, you can plan to explore important volcanic features at leisure before attempting to land on the plains and before venturing out onto the broiling surface to climb a mountain. Probably a special Venusian blimp will be available to you, to provide free, untethered travel from the floating station down through the lower atmosphere.

A good first goal for your airborne volcano study will be the Plains of Niobe. Named after an unfortunate Greek princess whose twelve children met a tragic end, the Plains of Niobe present a remarkable array of volcanoes to explore. As you slowly fly over them you will see nearly a thousand individual volcanic cones and domes. Most will be circular in outline but there are also some that are long and narrow, with linear vents at the top. Most rise above the surface like respectable volcanoes, but some are dominated by deep summit pits that lie below the surrounding plains level.

Over 200 good-sized shield volcanoes have been mapped in Niobe, based on Magellan radar images. Commonly, shield volcanoes on Venus are about 5 to 10 miles across at the base and have heights of a thousand

The Venus volcano Sif as mapped by Magellan radar measurements. The dark areas are probably similar to aa lava flows on Earth (NASA).

feet or less. As you fly over them you will notice that most have craters at their summits, some of which are quite small and others quite large, up to half the size of the whole volcano. These shield volcanoes are very like terrestrial ones and were probably formed similarly, by low-viscosity lava that issued from the central crater and flowed out freely to form the typical gently sloped mountain.

Niobe has other kinds of volcanoes, too. There are at least a couple of hundred cinder cones, steep-sided small volcanoes, probably formed by more viscous lava that was expelled more violently and cooled as small cindery blocks. These cones are small, usually less than 5 miles in diameter and almost all have clearly visible small craters at their summits. They seem to be very like the commonly-found cinder cones of the Earth, such as Sunset Crater in Arizona. Were you to decide to land and climb

A small volcanic crater in the California desert, similar in size and shape to the smaller cones found in the Plains of Niobe (author photo).

one, you'd probably experience the same slow progress as on Earth, where a climber slogs up through the loose cinders, sliding back part-way with each step.

Domes are also found in the Plains of Niobe. These volcanic features were probably formed by fairly high viscosity molten rock, more viscous than the freely-flowing basalt that is so common on Venus. The Venusian domes are not really very like terrestrial domes, however, as they tend to lack the expected widespread deposits of fine volcanic ash. This absence may be the result of Venus' dense atmosphere, which would suppress the formation of high-blown plumes of material. The Venusian domes are typically 10 to 20 miles in diameter, have flatish tops and steep-sided edges. Many have small summit pits and some are surrounded by radiating spurs and valleys.

Unlike anything on Earth are the so-called "pancake domes." Especially good examples are found elsewhere on Venus, for example in a neat cluster near Alpha Regio, an equatorial volcanic complex west of Niobe. The pancake domes there are about 15 miles in diameter, with steep sides, small central pits and highly fractured, nearly level upper surfaces.

A reconstruction of the volcano Sif on Venus, based on radar reflections. The vertical scale is highly exaggerated (JPL, NASA).

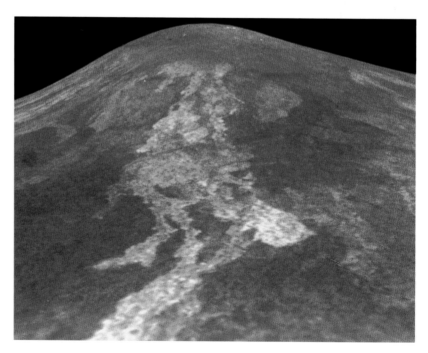

To the Bell tower of Venus

Having surveyed a typical "shield field" on Venus, you will want to prepare for your descent onto the surface and your attempt to climb one of Venus' large volcanoes. There are plenty to choose from. Radar maps have discovered over 400 good-sized volcanoes, with diameters ranging from 10 to 200 miles and summits rising to heights of 10,000 ft or more.

This chapter outlines a plan to visit one of these volcanoes, a peak located in an interesting area called Bell Regio. The name comes from English mythology, not the musical device. Bell was a female giant of ancient times. Appropriately, Bell Regio has a number of giant volcanic features, conveniently concentrated for your visit.

Before attempting to land your vehicle, it would be both wise and interesting to fly over some of Bell's giant volcanic features. The high volcano, Tepev Mountain, which we have somewhat misleadingly called a "tower" in the title of this section, will be left to last. Three other huge features, two of them volcanic, will be visited first by air. Bell Regio is a large, fairly rugged area, spanning some 700 miles, so it will take a couple of days' travel to explore its main highlights. You will want to fly low over these features, some of which are unique to Venus and still puzzling as to their nature and their origin.

The crown of Nefertiti

Named after a famously beautiful and influential queen of ancient Egypt, Nefertiti is an example of a giant Venusian crown or "corona." First clearly identified by the early Venera radar surveys, the features called coronae originally were quite baffling and they still generate some controversy about their true nature. Their properties were described earlier in this chapter and you will see that Nefertiti is fairly typical of the larger coronae. It is about 300 miles in diameter and has a rough raised outer rim over 1,000 feet high in places. Its central area is marked by flat plains, broken by hills that form an imperfect concentric inner ring.

Flying low over Nefertiti will give you a good chance to check out the ideas about how it formed. Is there evidence that lava ever broke the surface and flowed out to make the hills and rim or do the surface properties indicate that the crust was merely distorted from below, where magma (molten sub-surface rock) pushed up but did not ever succeed in breaking through? Most Venus geologists believe that coronae like Nefertiti were formed in the second way, without the presence of fresh lava flowing onto the surface. But Earth has nothing like Venus' coronae, so there is no local counterpart with which to compare them. Your flight could help solve the mystery, but it also will probably make you wish that there would be time for you to descend to the ground for a closer look and for collecting samples. Maybe at a later time, in a later book.

Craters and contrasts

Nefertiti lies near the northern edge of Bell Regio. After spending at least a day exploring the corona, your plan will be to fly west to see another interesting volcanic feature. Tipporah Patera is the name of a nicely shaped volcanic crater that is similar to volcanic calderas on Earth. There are about 100 large examples of this type of volcano, which can be thought of as having the wide summit crater of a volcanic mountain, but without the mountain. In shape a patera is rather like a large impact crater, with a nearly circular outline, a raised rim and a flatish center. But close inspection of the Venera and Magellan radar images reveals their volcanic nature; they have lava flows across their central floors and for some there are also lava streams flowing outward onto the surrounding plains.

Tipporah Patera is not very large compared with some of its sisters. Only about 50 miles across, it is, however, a giant compared with typical

Sacajewea, a volcanic caldera on Venus (NASA).

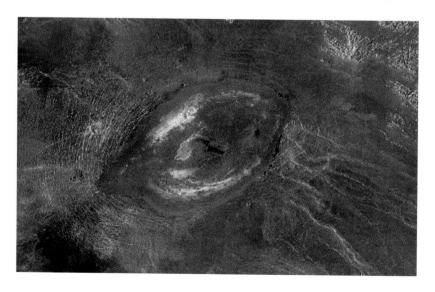

terrestrial calderas. In structure it is fairly similar to the Valles caldera on the flanks of which is nestled the famous atomic laboratory at Los Alamos in New Mexico. Both calderas are surrounded by fairly rugged country-side, which means in the case of Los Alamos pleasantly cool weather even in the summer. Tipporah, of course, never has pleasant weather.

Names on Venus provide a rich history of people on Earth. Thus, it is a joy to learn about the women who lent their names to Venus' landmarks, just as it is a joy to study those geological features. Tipporah Patera was named after the wife of Moses. Her name is variously transliterated as Tipporah, Zipporah, or Tsipporah and her fame is based on an incident that demonstrated her medical skill. She lived in the fifteenth century BC.

The next interesting feature for you to fly over also has an interesting name. Kottauer Patera is named after a famous fifteenth-century (this time AD, not BC) historian who lived and wrote in what is now Austria. The Kottauer Patera is larger and less circular in outline than Tipporah, so should provide you with some contrasts that may be useful and revealing. It lies about 200 miles southwest of the smaller crater, so it will take several hours to navigate your craft above the intervening ridged plains.

On the way, you should be sure to hover for a while over another kind of crater that is almost certainly not volcanic. The little circular feature called Piret (a common Estonian feminine first name) is most likely an impact crater. It is only about 20 miles in diameter and has a nice raised rim with some radar evidence of a surrounding narrow ring of rough ejected rubble. In size Piret is similar to the famous Australian impact structure called Gosses Bluff and to the Spanish impact feature at Azuara. But unlike these, Piret is not severely eroded like Gosses Bluff, of which only a central mountain ring remains, or almost completely covered by sediments like Azuara.

Piret will no doubt reveal important evidence to you about what happens when a natural missile collides with Venus. We know from the radar maps that Venus' thick atmosphere has a profound effect on the sizes and shapes of impact craters there. Few small impacts exist, because the smaller meteorites, comets and asteroids are completely "burned-up" (melted and evaporated) in the atmosphere. Some larger bodies are broken in two and these produce elliptical or double craters. Your close examination of Piret should give you some interesting facts to add to our still sparse knowledge about impacts on Venus.

Venus' Mayan monument

The Mayan people of pre-Columbian Meso-America were remarkable astronomers and magnificent architects. Their buildings and monuments are among the wonders of the New World. It is fitting that Bell Regio's highest, most conspicuous feature is a round, symmetrical mountain shaped like a huge but muted Mayan monument and named after

the Mayan creator goddess, Tepev. With a base of about 15 miles, Mt. Tepev rises above the plains to an altitude of nearly 14,000 ft, about as high above the Venus reference level as the Matterhorn rises above the Earth's sea level. But it is unlike the Matterhorn in both profile and origin. Mt. Tepev is a more gently sloped mountain, a shield volcano, typical of the hundreds of good sized volcanic peaks that are found all over the surface of Venus.

Mt. Tepev (sketch based on USGS maps).

The following paragraphs assume that your expedition will have available a set of high-tech environmental suits that can protect humans from the terrible heat of the Venusian surface. If this is possible, then you will be able to attain a first ascent; you will be the first party to climb a high mountain on Venus from base to summit, using your own power and your own well-protected legs. The feat will be more of an adventure than a desperately necessary scientific expedition, of course, as scientists will by then have collected many samples and close-up images of Venus volcanoes using robotic devices.

The plan will involve a landing onto the surface to the west of the mountain, at a point on the relatively flat plains at latitude 30°N, longitude 43°W. Care will be required in the landing because this portion of Bell Regio is not as smooth as some of the other plains of Venus, being marked by hills and ridges that are probably of both volcanic and tectonic origin. It would be worth a day of extra-vehicular exploration to determine the nature of the surface geology. This would also give you a chance to try out your Venus suit while close to the safety of the vehicle in case something goes wrong. It's important to remember what happened to the first spacecraft to land on Venus' broiling-hot surface: in spite of their heavily armored protection, they all died within about an hour of landing.

From the point where your vehicle landed and will wait for you, the straight-line distance to the mountain's summit is under 10 miles. Were there a nice trail and were you in comfortable T-shirt and hiking shorts, the climb would be an easy day trip for a hiker in good condition. However, with no trail blazed nor even a route determined, and with a heavy and complicated environmental suit to wear, you will find the trip to be a tremendously difficult physical challenge. And yet, it is not one that could easily be broken up into several day-long stages. The idea of lying down to sleep among the mountain's jagged, red-hot rocks is not a pleasant prospect. Instead it is probably better to plan to make the ascent in one long, exhausting day. The feat may not be more impossible than some of the spectacular twentieth-century climbs in the Himalayas, where skilled, nearly super-human climbers like Reinhold Messner scaled Earth's highest peaks solo and without oxygen.

The slopes of Mt. Tepev are fairly steep for a shield volcano. You will want to work out a route that stays with the smoother lava flows and avoids the more rugged areas. Our radar maps show both bright and dark

surfaces on Venus. The radar-bright areas are surfaces to avoid. Usually, brightly reflecting features mapped by radar reflection are surfaces that are rough on the scale of the radar wavelengths (inches or feet). Surfaces that look dark to radar, on the other hand, are comparatively smooth. They reflect radar back like a polished metal surface reflects visible light.

The Matterhorn, almost identical in elevation to Mt. Tepev (author photo).

Radar maps of Tepev indicate that there are both dark and bright areas on its slopes. These may be the Venus equivalent of what we find on the Hawaiian shield volcanoes. One of the two main types of rock that make up the slopes of Mauna Loa, for instance, are rough, rugged, sharp lava rocks called "aa." Hiking up an aa slope can tear a pair of street shoes to shreds in mere hours. Much more forgiving is the other kind of lava, called "pahoehoe." Its surface is smooth and shiny and travel across it can be as easy as crossing a paved parking lot. On a mountain slope, pahoehoe can be broken into blocks, with cracks and tilts making the going more of a challenge, but it is still usually easier to negotiate that than a steep slope of killer aa. So you should probably follow the radar-dark areas as you steam your way to the summit of Mt. Tepev.

When you are about 2 or 3 miles from the summit, the slopes will level off and you may be able to walk fairly easily to the volcano's highest point. Our radar maps are not detailed enough to tell in advance which of the rises that surround the summit crater is the highest, but your eyes can do that, and your climb will end there.

What will you see? How will the thick atmosphere distort your views? Some artists have tried to give an idea of Venus' dense air by drawing a weirdly distorted landscape. But it will probably not be too twisted. It will be a little like viewing the Earth's sea floor from a diving mask or glass-bottomed boat. Things will look a little closer than they are, but they should otherwise be in about the right place. Most of your view will be of the summit plains and the crater, but off at the horizon you may be able to make out in the dim Venusian light the rugged torrid plains of Bell Regio, to which you should hasten to return.

This trip is proposed as a long, strenuous day trip. Of course, we mean a terrestrial day, not a Venusian one. Because of its long period of rotation on its axis, daylight on Venus lasts about eight of our weeks. Your trip, including the days spent in the vehicle flying over Niobe and Bell Regio, can all be done during Venus' daylight hours.

Why, you might ask, don't we propose a plan to venture onto the surface in the cool night, instead, and then to climb the mountain by moonlight? Well, first, Venus has no moon. Second, the night is just about as hot as the day, as Venus' thick atmosphere and opaque cloud layers trap the heat and hold it in. But there might be a reason for visiting the surface at night. The hot surface rocks ought to give off a soft, red glow at night, so that the whole landscape should be visible as a ruddy, hellish fairyland beneath the darkened sky. Such a vista is bound to inspire any visitor with awe and wonder (and to precipitate a strong desire to return home to cool, green Earth).

11

The Cliff of Discovery

The planet Mercury is one of the nearest planets to Earth and yet it is one of the least known. Because it is usually so close to the Sun in the sky, few people have actually seen it from Earth. And as of the time of this writing, only one space mission has been sent to Mercury and that was something of a fluke, as you will learn below. Finally, that spacecraft only saw about half of Mercury's surface. The rest of the planet remains an unseen mystery.

The Moon or not the Moon?

One of the reasons that Mercury has been neglected is that it seems to look almost exactly like the Moon. If you should hold a spacecraft image of Mercury's surface next to an image of the Moon's, it would be difficult at first to figure out which was which. And yet, a longer look would reveal some important differences. You would see that Mercury is unique, a special planet with some strange and wonderful features. There are mysteries and controversies about it and, of course, there is that other half of its surface that remains a *terra incognita*, an unseen, unknown land.

Mercury is a small planet, though not the smallest (that's Pluto). It is larger than the Moon, with a diameter of 2,900 miles compared with the Moon's of only 2,100 miles. It's a heavy planet, with a density almost as large as the Earth's and much larger than the Moon's. This, in fact, is one of Mercury's big mysteries. We don't know why it is so dense. Apparently Mercury has an unusually large core of iron (at least, that's our best guess of what's inside to make it so heavy), but astronomers do not know why. Speculations about this peculiarity suggest that the answer may turn out to be really important — it may be a vital clue about how the inner Solar System formed.

Mercury's large density means that gravity on its surface is greater than on the Moon. The Apollo astronauts on the Moon, even with their heavy space suits on, felt light and bouncy. They could hop about the surface in a slightly comical way because of the Moon's low surface gravity. On Mercury, however, you will not have that fun. Its pull of gravity is twice the Moon's at its surface, and so, wearing your heavy space suit, you'll be more likely to stumble heavily about, wishing that the planet didn't have that giant iron core.

The reason that a first glance at Mercury reminds one of the Moon is that the surface is largely covered with impact craters. These craters show many of the characteristics of common lunar craters: deep centers, raised rims, central peaks in the large ones, and a surrounding blanket of ejected material. But the ejecta blankets, on closer inspection, look different. They tend to be less spread out than on the Moon, nestling close to their craters. We understand this difference; it's because of that higher gravity. When a meteorite hits the surface of Mercury, the explosive impact sends out a cloud of shattered rocks, but they don't travel far from their point of origin because the strong pull of gravity brings them right down close by, making a small, compact ejecta blanket.

The impact craters, nevertheless, look more like those on the Moon than the ones that we see on Venus or on Earth. The reason is that Mercury, like the Moon, has virtually no atmosphere to impede the travel of projectiles or to erode the craters as they age. There is an extremely tenuous atmosphere there, but it is made up just of a few particles of trapped solar wind and some atoms of surface material sputtered off the surface rocks by the impacts of the solar wind, cosmic rays and microme- teorites. This atmosphere doesn't amount to much. It is no more dense than a really good vacuum that we can create here in our Earth laborato- ries and it would not be good to breathe, as it is made up mostly of hydrogen, helium and heavier elements, like sodium and calcium.

Mercury's days, like the Moon's, are much longer than Earth's. Mercury's rotation period around its axis (which is lined up pretty much straight north and south) is just under 60 Earth days. This turns out to be exactly two-thirds of its period of revolution around the Sun; the Mercury year lasts 88 of our days. The reason for this apparent coincidence is one of the facts about Mercury that have intrigued astronomers since the 1960s. Somehow the Sun seems to have captured Mercury's rotation, apparently because of the intense tidal forces that it exerts, especially when Mercury is closest to the Sun. The planet has a rather elliptical orbit, so it sometimes is much closer to the Sun than at other times and it is the rhythm of this effect that seems to explain Mercury's peculiar rota- tion. The effect on the expedition is that you will have a good twelve weeks of daylight at a time, so that the mission will have plenty of light to

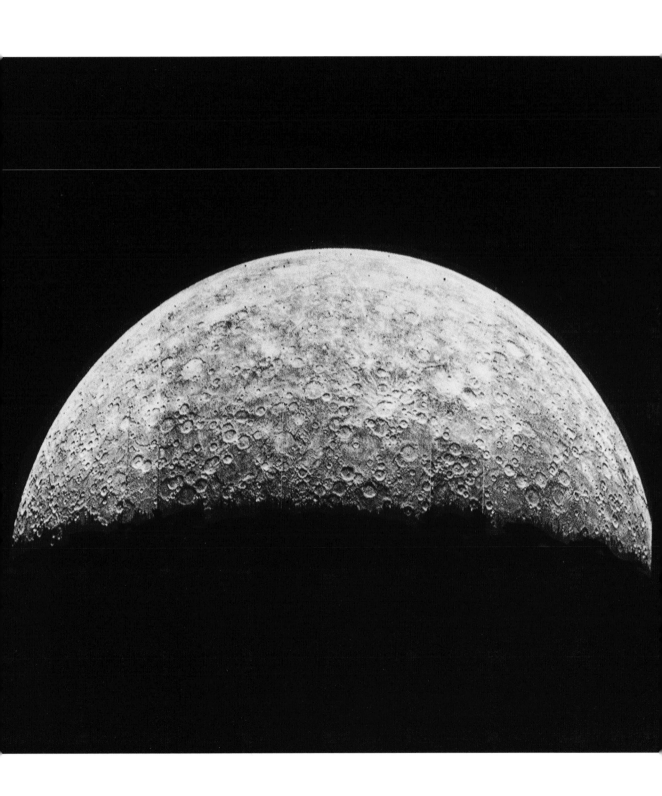

accomplish its tasks in a Mercury day. In this sense, Mercury is rather similar to the Moon.

An important difference, of course, is that Mercury is much hotter than the Moon. Its proximity to the sun explains why the temperature in the daytime is a sweltering 800 °F. As for Venus, the biggest challenge to explorers on Mercury is the intense heat. But otherwise, Mercury is a good planet for exploring, easy compared with Venus. There's no stifling atmosphere and the surface is relatively smooth and gentle.

The Kuiper caper

The bright crater named Kuiper, perched on top of the rim of an older, larger crater, named Murasaki (NASA).

We owe our current knowledge of the surface of Mercury to the efforts of one man. Gerard Kuiper was an astronomer whose passion was exploring

the planets. Born in the Netherlands, Kuiper spent most of his professional life in the United States, first at the Yerkes Observatory and later at the University of Arizona. Bright, industrious and imaginative, Kuiper accomplished an immense amount of good science, discovering new satellites, mapping the planets and our Moon, exploring the asteroids and studying the nature of cometary orbits. With the advent of the Space Age, he became a major figure in the planning of NASA's missions to the Moon and the planets, including the highly successful probe to Venus, Mariner 10. In the process of thinking about this mission, Kuiper did some calculations that led him to make a bold suggestion. He figured that, if the planning was done just right, Mariner 10 could be programmed to get two planets for the price of one. After passing close to Venus, where Mariner 10 would make many measurements and take the first good pictures of the cloud patterns, the spacecraft could be given an orbit that would then take it on to Mercury, without any substantial extra cost to the program. NASA agreed to the idea and our one and only mission to the planet Mercury was launched.

On November 3, 1973, Mariner 10 reached the innermost planet and we obtained our first glimpses of its cratered surface. On its way from Venus, Mariner 10 had been sending back images of Mercury that showed increasingly clear views, in which the impression of a lunar-like surface became clearer and clearer. Finally, at its nearest point to the planet, some 500 miles above the surface, it sent us wonderful images, revealing many different kinds of terrain and revealing similarities to the Moon, as well as some puzzling differences.

The first object that NASA scientists saw, well before the spacecraft's close approach, was a bright spot that appeared much whiter than the surrounding surface. Later, astronomers proposed that this initial discovery, the first surface feature to be distinguished, be named Kuiper as a memorial. Sadly, Gerard Kuiper did not live to see the fruits of his ingenious idea about Mariner 10. Before Mariner 10 reached Mercury he died of a heart attack while in Mexico, where he was involved in surveying sites for a new observatory there.

The object named Kuiper turned out to be an impact crater that has a higher reflectivity than its surroundings, probably because of its relative youth. It has not yet been subjected to the ages of erosion of the surface caused by meteorites and solar wind particles. It is about 40 miles across, a medium-sized crater not distinguished by any special physical charac-

The giant Caloris Basin, Mercury's version of a mare basin (NASA mosaic of Mariner 10 images).

teristics other than its youth. But a man-given characteristic does significantly distinguish the crater Kuiper. It's the only crater not named after a person in the arts.

As images of the surface of Mercury began to come back from Mariner 10, astronomers saw that it is covered with thousands of impact craters, as well as a few other types of features. The task of naming these objects fell to the nomenclature commission of the International Astronomical Union. Feeling that the names of craters on the Moon had sufficiently honored scientists of the past, the commission decided that Mercury should be a place that would commemorate the great artistic people of the Earth. Therefore all of Mercury's named craters (with the exception of Kuiper) were given names of painters, musicians, writers and other great figures in the arts. For example, there is the giant crater Bach, 135 miles across, the crater Van Gogh, about 60 miles across, and the large, ancient crater Homer, some 200 miles across and lying just to the west of the smaller and much younger crater Kuiper.

The riddle of the rupes

In spite of a few differences, the craters on Mercury look a lot like the Moon's. But there is another type of prominent feature of this planet that is very different from anything found on our satellite. These are immense cliffs, called "rupes" from the Latin word. Found here and there all over the surface, these cliffs are long, usually curved and in some cases quite steep, compared with other features on Mercury. Most of them seem to be fairly young, as they cut right through other things, like craters and plains. Without samples of rocks from Mercury, we don't know exactly how old its different features are, but crater counts give an approximate age, and we find that most of the larger craters, like Puschkin and Puccini, date back to the period of catastrophic bombardment, when the lunar basins were formed and the Solar System was cleaning out its store of large interplanetary debris, about 4 billion years ago. If so, then the giant cliffs must have been formed a little more recently, perhaps 3.5 billion years ago or so. Something happened then to cause these widespread scarps to appear.

When first glimpsed by terrestrial scientists, they were a mystery. The cliffs seemed to be the result of large-scale crustal movement, like that caused on other planets by the creation of faults when one piece of a planet's skin breaks and shifts position with respect to an adjacent piece. However, the Mercury cliffs were seen to be quite different from the effects of shifting we find on the Earth, caused by earthquakes and plate tectonics. Without involving lateral shifts, where a piece of surface moves sideways or where two sections pull away from each other, Mercury's faulting was only of one type, called "thrust-faulting." One piece of land was thrust up with respect to the land around it.

The mystery of the rupes must be connected in some way to some other unique properties of this planet; there must be some physical reason why it experienced faulting of this particular kind, all over its surface, all at about the same time. The most reasonable suggestion that people have is that the rupes are somehow related to the fact that the interior of Mercury is fundamentally different from that of the Moon. The Moon has a very small central core of heavy material (iron or iron compounds), while Mercury's large density indicates that it has a giant iron core, making up about 75% of the planet's radius. Apparently after the planet cooled, following its early violent formation period, its crust solidified first, preserving the cratering record of the period of bombardment. Its huge interior of iron cooled later, shrinking as it solidified. This shrinking is thought to be the cause of the giant cliffs. Exactly why they formed in the shapes that we find them, with the sizes that we see, is not really understood. Your expedition to one of these marvels should help to clear up the riddle of the rupes.

The Discovery Dome dispute

The goal of this journey to the Inner Planet is the spectacular Discovery Rupes. Almost 300 miles long, Discovery Rupes is one of the biggest of Mercury's weird cliffs. Unlike the impact craters, which are named after people of the arts, the cliffs of Mercury bear the names of famous ships of exploration. Discovery Rupes, for instance, is named after one of Captain James Cook's ships, and there are others, such as Endeavor, named after another of Captain Cook's ships, and Santa Maria, after one of Columbus' ships.

The scarp named Discovery lies in the planet's southern hemisphere at a latitude of 55 ° S and a longitude of 40 ° W in a rugged area south of the giant crater Schubert. It is curved in shape, with two gentle scallops. There are smooth plains towards the north at the bottom of the cliff,

The Cliff of Discovery (Discovery Rupes), photographed by Mariner 10 (NASA).

while the uplands leading away from the top of the cliff are heavily cratered in the north and smooth at the south. The smooth areas are typical of the terrain called "intercrater plains," Mercury's poor excuse for maria like the Moon's. Such plains are found scattered over the half of the planet that we've seen and they all seem to have the same age. Possibly they represent a period in Mercury's history, after most of the early cratering, but before the formation of the scarps, when there was global volcanic activity, with much of the surface, here and there, covered over with lava. Alternatively they may be just smooth plains of impact debris like the Apollo 15 astronauts found on the Moon.

When the Cliff of Discovery formed, it divided the small plain in two, separating the pieces by a vertical difference of thousands of feet. A few craters were also divided up, telling us the sequence of events: first the initial bombardment, then the volcanism that formed the plains, then some more cratering, and finally the shrinking of the core and the formation of the giant cliffs.

If we turn back to the period of volcanism for a moment, the question arises, how did the lava arrive at the surface? Did it issue from cracks in the surface or did it flow out of volcanic mountains? Probably such cracks of origin would not be visible in our images, but a good volcano ought to be a conspicuous feature on our maps. Or should it? Anything like the giant volcanoes of Mars would surely have been seen, but small volcanoes like those so common on Venus or the Earth would not be so obvious. Perhaps, it has been argued, there are small volcanic mountains scattered over the surface but almost invisible to Mariner 10's cameras. Scientists are sharply divided on this question. Some point to certain rather fuzzy images and claim that they're volcanoes, while most argue that the images do not support such a claim. Your expedition, among its various adventures, will have an opportunity to resolve this dispute.

One of the possible volcanoes lies right on top of Discovery Rupes. Called the Discovery Dome, this feature appears to be a small (perhaps 10 miles across) pimple just west of the lip of Discovery Rupes, near its center. At the top of the dome there is a small depression that may be the crater of this possible volcano. The gentle slopes that fall away from the crater are rather like those of a small shield volcano. On the other hand, the resolution of the Mariner 10 pictures is so poor that it may not be a separate mountain at all, but just a remnant of the rim of an old, eroded crater or possibly a large block of ejecta (there are similar features, called the knobby deposits of the Odin Formation, that are identified as blocks of ejecta from the immense collision that formed the Caloris Basin). Certainly, inspection of the feature from the ground, right there at its foot, will settle the dispute.

Rameau and its ramparts

Just about in the middle of the Cliff of Discovery there is a good-sized crater called Rameau. This nicely shaped impact feature is named after Jean-Phillipe Rameau, who was a seventeenth-century French composer, remembered especially for his elegant harpsichord music and his operas.

The crater Rameau is about 40 miles across, with steep sides leading down several thousand feet to an almost flat floor. The remarkable thing about Rameau is that the Discovery fault passes directly through it, cutting it neatly in half. The western half of Rameau lies high above its eastern half. This shows, of course, that Rameau is older than the cliff. In

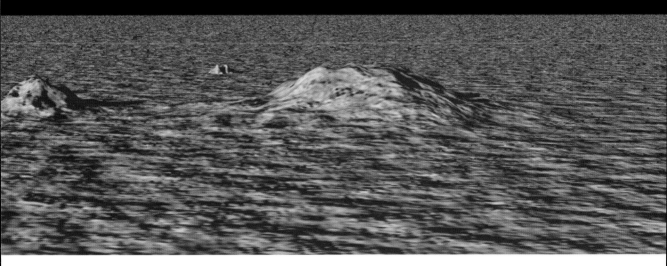

the history of this part of Mercury, after the period of volcanism that formed the intercrater plains, a large asteroid or comet, some 5 miles across, collided with the planet, forming the crater, and it was later that the crust split and the huge cliff formed.

The scarp cuts its way jaggedly through the flat floor of the crater. For most of its length it is a steep, high cliff with its top rim rounded a little by erosion and avalanches. Near the northern end there appears to have been a major landslide, with the rock forming a huge talus slope that extends out onto the crater floor about 2 miles from the foot of the cliff. Near the south wall of the crater the cliff has collapsed so much that it appears to have left a veritable highway that leads gently up from bottom to top. Perhaps some day an exploring crew will use this tempting route to ascend the Cliff of Discovery. But not your crew, who are scheduled to undertake a more challenging venture outside of Rameau's ramparts.

Climbing the Cliff of Discovery

Your arrival at Mercury will at first be reminiscent of that wonderful pioneering venture, the lunar Apollo mission. After parking in orbit for a

An imaginary view of possible volcanic domes on Mercury. The lack of an appreciable atmosphere means that the sky is black even in the daytime.

few days (terrestrial days, of course, not Mercury days), the time will come for the Mercury lander to separate from the mother ship and descend to the hot desert plains below. The landing spot will be strategically placed to the east of a section of the scarp just south of Rameau, where the cliff appears to be best-preserved, steepest and least eroded by collapsing land-slides. This is not a perfectly smooth landing place, so care will be needed in selecting the right place to settle the spacecraft onto the surface. There appear to be numerous small craters and some ejecta, largely from Rameau and from a smaller, unnamed crater to the southeast. But with the experience of the Apollo landings behind us, we know how to select a comfortable landing site, so there should be no problem settling gently onto the planet's surface.

Stepping out of your spacecraft will be a bit of a challenge. If some of the crew have by now had experience with travel on Venus' surface, Mercury will seem fairly easy. Though almost as hot, the gravity will be sufficiently less to make moving about with your bulky space suit a little

A sketch of the Cliff of Discovery as seen from the surface below it to the east.

more pleasant. You and your suit will weigh a little more than a third of your weight on Earth. Figuring about 100 pounds for the protective clothing and its built-in supply pack, together you will feel as if you weigh only some 80 or 90 pounds, which ought to give you a certain optimistic bounce to your steps as you begin your exploration of the landscape around the landing place.

Of course, most of your suit is devoted to providing you with air for breathing and refrigeration for keeping cool. The surface temperature in the Mercury daytime is typically about 800 °F. This might tempt your expedition leader to consider landing in the Mercury night and exploring by searchlight, but the night-time temperature would be so cold, about −280 °F, as to make that plan just as challenging, so probably it will be best to continue with the plan to use the advantage of the clear daytime sunlight.

We do not have any intimate knowledge of the roughness of the terrain that you will face on your journey across the surface to the base of the Cliff of Discovery. Radar indicates that the going will probably be pretty rough. Small-scale cratering has probably left a tangle of pits and blocks of rocks that will require careful route-finding, even across what appears to be a pretty smooth surface on the Mariner 10 images. Because of Mercury's greater mass than the Moon's, meteorites that hit the surface do so with a much greater velocity, being pulled in to the surface with a higher gravitational pull. This probably results in steeper, sharper and larger craters than would have been formed by the same projectile on the Moon. So you should be prepared for a rugged journey.

The point to which you will be heading is at the foot of the cliff exactly 10 miles south of the rim of Rameau. You should probably take a Lunar Rover-type vehicle there, to give you plenty of time and energy to make the steep ascent. A full day could be spent exploring the base of the cliff, doing the geological research that should illuminate the process that formed this huge feature. But soon you will be ready for the main event – the climbing of a fabulous and unique planetary wall.

We know that it is pretty steep, averaging about 30° or so, though, as in any mountain slope, there will surely be many places that are much steeper. Though it looks from the images as if it is almost completely vertical, the actual distance from foot to rim is about 4 miles as the crow flies (though, of course, no crows fly in Mercury's almost imperceptible atmosphere). From bottom to top the vertical distance is about 5,000 feet, not unlike the depth of Earth's Grand Canyon. In fact, the trip will be rather similar to hiking up one of the trails of Grand Canyon National Park from river to rim, except that there is no atmosphere, the temperature is many times hotter, there is no river, there is no comfortable hotel at the top, and the cliff was formed by a different process. A pleasant difference will be that there will not be any donkey trains full of tourists to force you off the trail, though, of course, there also will be no trail.

This trip will be a pioneering effort and pioneers always have to be prepared for surprises. Instead of nice, climbable rock, it is possible that the cliff will be made up of hazardous scree-like pebbles and sand, being covered with a layer of impact debris that will be hard slogging. You will want to be sure that your boots are made with textured soles that can grip loose and unstable soil.

To allow enough time for the geological research and rock collecting that you will want to do, you should plan a full day for the ascent and another day to descend, with a rest period spent on the rim. It would be interesting to return down to the Rover by an alternative route, just north of the steep ascended slope. There appears to be a small crater about halfway down the cliff. If it is an impact crater, it may provide you with some interesting information about the cliff walls, as the impact will have excavated sub-surface rocks that will be fresher than those that you walked on going up.

Finally, after three Earth days of exploration, the exploring party will be ready to return to the lander and eventually to the mother ship in orbit. Your cameras will be full of images and your sample bags full of rocks, all illuminating the true nature of one of Mercury's mysteries, the remarkable Cliff of Discovery.

12

Descent into the maelstrom

Jupiter is the largest planet. Its diameter is ten times the Earth's and its mass is so large that it dominates the gravitational activity in the outer Solar System. It can capture comets, perturb the motions of the nearby planets and influence the orbits of the asteroids. It even has the reputation of having flung objects into the inner Solar System, where they have collided with the Earth.

It is big and imposing but, as a planet for exploration, Jupiter leaves a lot to be desired. It has no solid surface. What we see from Earth, or even from visiting spacecraft like Voyagers 1 and 2 and Galileo, is just the top of

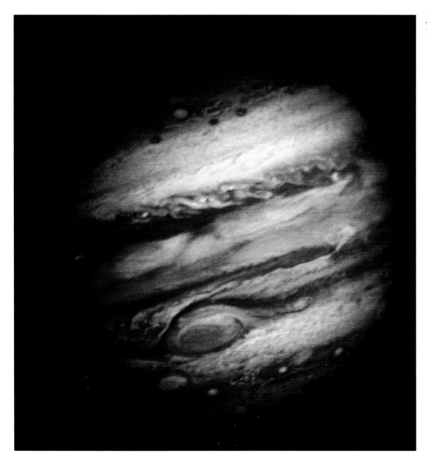

Jupiter as photographed by Voyager 1 (NASA).

an immense cloud layer. Below these icy clouds are more clouds, and beneath them the atmosphere gets warmer and thicker until it gradually becomes so thick that it turns into a liquid. At great depths the material of Jupiter's interior, mostly hydrogen, is squeezed together with such high pressure that it becomes a weird kind of stuff called metallic hydrogen, a form of that element that only can exist at the extreme conditions found in a place like the insides of a giant planet. There probably is a rocky core at the center of Jupiter, but we won't ever stand on its surface, if it has such a thing. Because its great mass pushes in so hard, the pressures there are immense and the temperatures are as high as 50,000 °F. If Jupiter were just a few times larger, its core would be hot enough to ignite thermonuclear reactions. Jupiter would be a star like the Sun and we would be living in a double star system.

Into orbit

Though it won't be possible ever to land on Jupiter's surface, there is little to prevent us from trying to explore it with a properly prepared craft of some kind. Of course, there will be many dangers. One is the intense radiation field that envelopes Jupiter and its inner moons. Jupiter has an immense magnetic field that has captured charged particles, making a huge, dangerous cloud through which you will have to tunnel. Plans to protect you from too high a dose of radiation will have been made by your vessel's engineers, increasing the complexity of your trip. But even to get to Jupiter is a major challenge. Manned spacecraft to the inner planets require months of boring travel in empty interplanetary space, but Jupiter and the other outer planets are much farther away and a trip to them takes years. The separation between the planetary orbits increases outwards, so that each step to the next planet is bigger than the last. Mars is only about 50 million miles from Earth at its closest, while the jump from Mars to Jupiter is over 300 million miles. And the next planet, Saturn, is another 400 million miles past Jupiter. So the first problem to solve on a flight to these outer planets is how to survive the two or more years it takes to get there. You can plan to pass by some interesting asteroids along the way, of course, which will relieve some of the monotony, but it still will be a long, long flight.

When you do arrive, though, it should be a tremendous thrill. Jupiter's immensity and its colorful clouds will fill your spacecraft's window in spectacular brilliance. If you plan to go first into an equatorial orbit around the planet, you will be close to its orbital plane and not far from the plane of Earth's orbit, as well. All of the planets revolve around the Sun in orbits that very nearly share the same flat zone of space. Jupiter's orbital plane is only 1° different from Earth's. Only Pluto's unusual orbit departs from this zone by more than 10°.

As your spacecraft orbits rapidly around the giant planet's equator, you will have a chance to survey the clouds below you, seeing their structure and motions up close for the first time. But the details will whiz past at a dizzying speed. Jupiter is so massive that its gravitational pull forces you into an orbit that circumnavigates the planet in about 3 hours. Your speed will be almost 100,000 miles per hour.

Bands and belts

Spread out below will be Jupiter's colorful clouds. The basic pattern of these clouds is much more permanent than that of the clouds on Earth and they cover the whole planet from pole to pole. For as long as astronomers have mapped them with their telescopes, beginning back in the seventeenth century, the clouds have looked pretty much the same. Even a small telescope from Earth will show the pattern of bright and dark bands, which line up parallel with the equator and which change little with time. The bright zones are called bands and the dark lines between the bands are called belts. Each is roughly 20,000 miles wide from its northern edge to its southern, so they occupy a range of latitude that is wider than the entire Earth.

The bands and belts are fairly stable in their location, but they can vary a lot in their detailed structures, as you might expect, considering that they are made of clouds. They can change in brightness or color over time and they can have temporary blemishes – white or brown spots – that can suddenly appear, only to fade away weeks, months or years later. Only one such spot, called the Great Red Spot, has shown any tendency to be at all permanent. The Great Red Spot has no counterpart anywhere else in the Solar System and it is this remarkable feature that your expedition will explore.

As you speed by in orbit, your view of the detail in the belts and bands will include something that we can't see from Earth: the three-dimensional structure of the cloud tops.

Twentieth-century spacecraft to Jupiter (Voyagers 1 and 2 and Galileo) showed us that the bright bands are higher in the atmosphere than the darker belts. It's the height of the clouds that determines how bright they are, as the high ones receive more sunlight.

We've known for many decades that the visible clouds of Jupiter are not made of water or ice like those of the Earth. Instead, analysis of the light reflected from them shows that the upper layer of clouds is made up of ammonia ice crystals. The temperature is cold enough (about $-180\,°F$) that the small amount of ammonia found in Jupiter's atmosphere freezes out, forming a thick layer of ice clouds. The gas in which these crystals float is mostly hydrogen, as that is the element that makes up over three-quarters of the mass of Jupiter, but there is also, at least in the part of the atmosphere that we've sampled, some helium, a little ammonia and some methane.

The visible cloud layer of Jupiter is just the top of a series of cloud decks. The next one down has been glimpsed and we have found that it is warmer than the top layer (about $-100\,°F$) and made up of a different substance, crystals of solid ammonium hydrosulfide, a chemical compound made up of hydrogen, nitrogen and sulfur. Below this layer, some 15 or 20 miles lower, is thought to be a thick zone of clouds made of water ice at a temperature close to $30\,°F$. Beneath that there may be additional layers, perhaps made of liquid water droplets. Our understanding of these deeper layers is still pretty tentative, as the Galileo probe, described below, failed to confirm some of our expectations.

To and fro

The cloud surface of Jupiter rotates much faster than the Earth's surface. Jupiter turns on its axis quite rapidly, with a period of just under 10 hours. This kind of rapid rotation is a characteristic of the giant planets out beyond Mars. While Earth and its small, rocky neighbor planets rotate slowly, the big outer planets all turn in just a fraction of an Earth day. This interesting difference is just the opposite of what you might expect: shouldn't the huge massive ones move more slowly and ponderously? The answer would be yes if all the planets were given the same amount of angular momentum (which is defined as the amount of mass and the amount of turning motion) when they were formed. Clearly, they were not, and this remarkable difference is an important clue to the mystery of the formation and the early history of the planets.

When we watch the equatorial clouds of Jupiter rotate from a relatively stationary place, such as the Earth, they are moving around the planet at speeds of about 27,000 miles per hour. The speed is slower, of course, at other latitudes, as there is less distance to travel closer to the poles. Actually, the speed also gets slower for another reason: the period of Jupiter's visible surface is not the same at all latitudes, as it would be if it were a solid surface. While the period of rotation at the equator is 9 hours 50 minutes, the clouds at intermediate latitudes, around 45° north or south, take 9 hours 55 minutes. Near the poles, the period is even longer. You can imagine what it would be like if the Earth's surface did this. Locations would be continually sliding past each other and your neighborhood, except for the houses exactly east and west of yours, would be continuously changing. Obviously, this kind of thing can only happen on a planet if the surface is a fluid.

Jupiter with two of its satellites seen in front of it. Io is at left and Europa is at right (NASA).

For Jupiter's clouds there is a further remarkable fact. The two Voyager spacecraft found that there are fierce winds in the planet's atmosphere, ranging up to 200 miles per hour and higher. The winds alternate as one looks at different latitudes, changing from eastward to westward, varying as one goes from the bright belts to the dark bands. These winds extend down deep into the lower layers of the atmosphere, as shown by the spectacular trip of the Galileo probe.

Galileo's plunge

Most of what we know about Jupiter's atmosphere below the cloud tops comes from an experiment carried out in 1995. The spacecraft Galileo, launched from Earth in 1989, reached the giant planet six years later, having had to acquire enough energy for the trip by passing close to Venus (once) and to the Earth (twice) to get some gravitational assistance. When it reached Jupiter, it released an instrumented probe that plunged down into the atmosphere. It descended very rapidly at first, but then was slowed when a drogue parachute opened up. This caused it to lose most of its sidewise velocity. Its main parachute opened up about 3 minutes after it entered the atmosphere at a position about 10 miles above the cloud layer. The probe then settled into the increasingly thick atmosphere, reaching the clouds about 8 minutes after entry.

This part of the trip was not an easy one. Jupiter's large mass means that things happen fast near it. The speed of the probe when it entered the atmosphere was over 100,000 miles per hour. When it was decelerating in the upper atmosphere it suffered almost incredible forces, withstanding about 200 g's (200 times the gravity we feel here at the surface of the Earth). As it screamed into Jupiter's hydrogen gas, the shock layer at its nose heated up to 25,000 °F, burning away much of a carefully prepared heat shield.

The probe's instruments sent back weather reports as it went down. The atmospheric pressure at the cloud layers was found to be just a bit higher than one bar (the atmospheric pressure here at the surface of the Earth). Winds were very high, faster than expected, reaching speeds of almost 400 miles per hour. This is faster than the highest winds experienced in terrestrial hurricanes and tornadoes. The probe found that these high winds extend down deep into the lower regions of the atmosphere, indicating that they are caused more by Jupiter's interior heat and circu-

[opposite] A sketch showing the Galileo probe descending through the cloud layers in Jupiter's atmosphere.

lation than by the heating from above by the Sun, as in the case of Earth's weather.

Temperatures measured by the probe were pretty extreme, too. At heights of about 200 miles above the clouds the temperatures reached the surprisingly high value of 1,600 °F. Lower down, at heights of 10 miles or so, the temperature dropped to a chilly −200 °F. Below the clouds, it rose again. When the probe was finally destroyed by the extreme conditions found some 80 miles below the cloud tops, the temperature was about 300 °F and rising.

The death of a comet

Galileo's probe was not the first thing to be eaten by Jupiter's mammoth gravitational appetite. Since the giant planet was formed it has swallowed up anything that came too close to its pull. Astronomers have known for a century or more that comets and asteroids with orbits that bring them within Jupiter's range can have their paths changed into much smaller orbits that are dominated by Jupiter. Some of the outer moons of Jupiter appear to be such captured objects. But in 1994 the world witnessed a spectacular event that showed just how powerful Jupiter's great mass can be.

The story began in 1992 when three astronomers discovered a remarkable-looking comet out beyond Mars. Caroline Shoemaker, who has discovered more comets than any other living person in the world, saw it first and was amazed by the fact that it seemed to be a linear object, rather than a fuzzy dot like most comets when they are that far from the Sun (before their tails have formed). The two other astronomers in the telescope dome at Palomar Mountain, Eugene Shoemaker and David Levy, peered at the discovery photograph and agreed that it was not a normal-looking comet. As they and other astronomers, alerted by the comet's discovery announcement, followed its path in the sky another remarkable fact turned up. Instead of orbiting the Sun, Comet Shoemaker–Levy 9 (so-called because it was the ninth comet to be discovered by the team) was orbiting around Jupiter! By extrapolating back in time, it was found that this object had ventured too close to Jupiter, which gravitationally captured it some time before Caroline Shoemaker first saw it.

When this funny-looking comet was examined with larger telescopes, such as the Hubble Space Telescope, it was found that its unusual appearance was the result of its being a string of about 20 smaller objects, all following each other in a compact line around Jupiter. Clearly what must have happened is that the comet, when it got within Jupiter's grasp, was torn apart by Jupiter's strong tidal forces. As more observations came in, astronomers calculated an increasingly accurate orbit and soon came to the realization that the comet was doomed. When it would next reach the point in its orbit near Jupiter, the comet's path would be fatal. It would plunge into Jupiter in a spectacular series of collisions.

The world waited anxiously for the big event, scheduled by nature to occur in July 1994. Right on time, one by one, Comet Shoemaker–Levy 9's pieces slammed into Jupiter in a spectacle that was witnessed by millions of people, many through their own small telescopes. Each fragment plunged into the giant planet's atmosphere with a velocity of about 130,000 miles per hour. The pieces were about a mile or so in diameter and each collision packed as much destructive force as a million atom bombs.

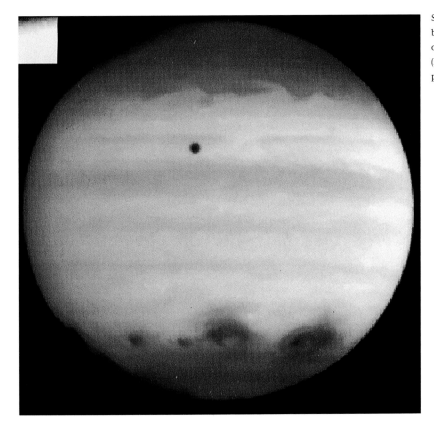

Scars in Jupiter's clouds (at bottom) were left by the impacts of Comet Shoemaker–Levy 9 (NASA, Hubble Space Telescope photo).

It took nearly a week before all of the comet's fragments were gone. It happened that the point of impact was just over the horizon, on the early morning side of Jupiter. The effects of the blasts, however, were so immense that large telescopes could see the plumes of hot material the instant that they rose up into the atmosphere above the horizon. And Jupiter's rapid rotation brought the scars into view only a few minutes later, so that we on Earth could see what happens when you get too close to this voracious planet. It will be an important lesson for the engineers who will be designing the vehicle that will carry you more gently (we hope) down into Jupiter's hostile environment.

The Great Red Spot

Giant storms on Earth, tornadoes and hurricanes that cause so much damage on the ground, commonly rotate like huge pinwheels. This rotation is caused by what's called the differential rotation of the planet's surface. At the equator, the Earth's surface moves about 1,000 miles per hour because of the planet's rotation, but at other latitudes, the velocity is less, as, of course, those portions of the surface have less far to go in a day. A large cell of air, therefore, will have a higher velocity on its side towards the equator than on its other side. Under the right conditions this can lead to rotation.

Jupiter has an immense storm in its atmosphere that has been seen for almost 200 years. Called the Great Red Spot for its vivid color, this long-lived storm lies at a middle latitude in the southern hemisphere. It rotates counterclockwise with a period of about 6 Earth days. Around its borders the Jovian clouds are pulled around with it, dragged by the Great Red Spot's rapid rotation. It is oval in shape, with its long axis stretched out along its latitude line, and in size it is huge, with an area big enough for two Earths to fall into it. The Galileo spacecraft showed that it is a high-pressure area, a tower of rising clouds, levitating up from the hotter depths of the atmosphere as it rotates. The bright red and white colors in the central parts are the highest, while the darker clouds in the surroundings lie lower down.

The bright colors of Jupiter's atmospheric clouds remain something of a mystery. The Great Red Spot's color seems to fade and intensify over time, as its pulls up different materials from below. The bright color must come from some compound, probably involving sulfur or phosphorus,

that is formed by Jupiter's internal heat from below or, perhaps, sunlight from above. The nature of the redness of the Great Red Spot is one of the puzzles that will be solved by your descent into this maelstrom.

The Great Red Spot as seen by Voyager 2 (NASA).

Descent

Astronomers have dreamed for years about exploring Jupiter's atmosphere, but the formidable difficulties of getting there have seemed almost insurmountable. Once there, things may not be so bad. A well-protected gondola hanging from a sturdy balloon can float comfortably in the clouds, allowing a thorough exploration of their chemistry and their motions. If you proceed with this journey, the good news is that the view from the window will be spectacular and the scientific data that you will collect will be extensive and decisive.

The bad news is that the descent will be extremely difficult and dangerous (and there is further bad news that we will not discuss now, but rather at the end of this chapter). The problem with the descent is that Jupiter's great mass makes for horribly high speeds and your spacecraft will have a very difficult time going from orbit outside the atmosphere to a position at rest with respect to the clouds inside the atmosphere. When the Space Shuttles or other manned spacecraft reenter the Earth's atmosphere, they withstand tremendous heat as they first collide with the thin upper atmospheric gasses. Special heat-resistant tiles were developed for the Shuttles so that they can come back safely, and yet these tiles can be so severely damaged by re-entry that some have to be replaced for each flight. In the case of Jupiter, the entry

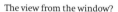
The view from the window?

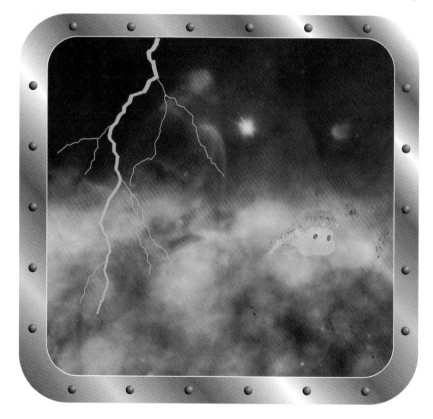

into the atmosphere is much more rapid and your craft will need to be very well insulated.

Atmospheric entry provides other problems, too. You will not want to plunge down into the atmosphere as precipitously as the Galileo probe did, as it had to withstand a deceleration of over 200 g's, far more than Shuttle astronauts put up with and too much for human survival. You will need to do things much more gradually and, fortunately, the engineers will have worked out how that can be done without using too much of your precious rocket power. Perhaps they will plan a series of "skips" in which your craft dips down into the atmosphere and bounces back out again, allowing a more gradual trip.

Once comfortably in the atmosphere, there will be many things to look for, but two are of special interest. The first is both scientifically important and personally dangerous: lightning. We know that lightning storms are common in Jupiter's atmosphere, as we have seen the strokes from space probes and we can hear their radio signals with Earth-bound radio telescopes. You will not want to have your temporary home zapped by a Jupiter-sized lightning bolt and it will take some careful planning to know how to avoid likely spots for this kind of stormy activity. The Great Red Spot, in spite of the fact that it presents some intriguing mysteries, may not be the safest locale. The Galileo orbiter saw some huge thunderheads in its vicinity.

The second special phenomenon to look for is less likely to be there. Scientists and science fiction writers occasionally have speculated that Jupiter's atmosphere, considering its low-depth warmth and the probable presence of various organic molecules, might be inhabited. Perhaps giant, gaseous creatures float about among the clouds, evolving and living in a way that is entirely foreign to what we consider normal for life forms. No one thinks that such a thing is very probable, but nevertheless, it will be interesting for you to be alert to the possibility that some of those billowing things outside your window might be in some sense alive. Long ago when the first Viking spacecraft landed on the surface of Mars, the astronomer Carl Sagan said that the first step in the search for evidence of life would be to look out through the camera's lens at the landscape to see if anything is walking past. You should plan to check Jupiter out in the same spirit.

Now we come to some more bad news. It was difficult enough getting down into Jupiter's atmosphere without burning up, but it is also quite a problem to get back out. It will be necessary to leave much of your weight there – the gondola, the various pieces of scientific equipment and other heavy materials. Most of the valuable data that you collected will have been radioed up to the orbiter, so that nothing vital will be lost. Only a lightweight cab, carrying bare essentials, will get you back up, as your vehicle must fight against a pull of gravity that is nearly three times that at the Earth's surface. Though it will be sad to have to leave so much behind, you will have turned on a radio beacon so that future visitors can, if they want, revisit your small Jupiter balloon-borne explorer, if it has not been destroyed by Jupiter's storms. Centuries from now, space historians may want to study it to learn about your pioneering journey, the first to descend into the maelstrom.

13

An Ionian adventure

On Earth an Ionian vacation is likely to be a relaxing and civilized affair. The Ionian Sea lies off the west coast of Greece, south of the Adriatic. It is famous for its lush, scenic islands, the best-known of which is beautiful Corfu at the northern end of the island chain. The sea and the islands were named after the beautiful young girl, Io, one of the objects of Zeus' liberally bestowed affections. According to the legend, Hera, Zeus' wife, became jealous of Io and, to save her from Hera's rage, Zeus changed Io into a cow. But Hera was not fooled and she made Zeus give her the cow as a present so that she could keep the poor creature under her watchful eye. Eventually Io escaped, only to be pursued by a horrible biting fly, known as a gadfly. Chased by the fly, she fled to the far reaches of the world, including the beautiful islands that now honor her name. Thousands of years later, cows continue to be annoyed by flies.

Tourists to the Ionian Islands are more likely to be coddled than annoyed. The seven islands are among the most beautiful places in Europe. The Ionian adventure of this chapter is also involved with an exotic and colorful place. However, the planetary Io is nowhere nearly as hospitable as a Greek island and your trip to this remarkable place will be a real challenge as well as a thrill.

Galileo's four

The astronomical Io is one of four remarkable discoveries made by Galileo that overturned the scientific world. When he turned his telescope to Jupiter in 1610, he was rewarded by an amazing sight: four tiny "stars" were lined up around it. As he watched these little Jovian companions from night to night, he realized that they were revolving around Jupiter. Galileo correctly surmised that they were moons like our Moon and realized the importance of this discovery: *contrary to doctrine of the times, not all bodies in the universe revolve around the Earth.* It had been believed by most people that the Earth (naturally, as it's our home) must be the center of the universe and thus everything must revolve around it. It's not a strange thing to believe. Standing on the Earth, we do not feel any motion. The inhabitants of the heavens, the Sun, Moon, planets and stars, do seem to

Jupiter and the four Galilean satellites as seen in a small telescope from the Earth. Io is at the right, near the edge of Jupiter's image, while at the left is Europa (barely visible, Ganymede and Callisto, from right to left (University of Washington Observatory photo).

be moving across the sky each day and night. It was not just egotism that led to the concept of an Earth-centered universe. But it was egotism that proclaimed that everything *must* revolve around the Earth, a tenet of many classical Greek philosophers that was adopted by prominent European thinkers, most of whom were the officers of the Church.

Galileo's proclamation that the four moons of Jupiter proved that not all things revolve around the Earth got him into some serious trouble with the Church. Although it had been suggested a century before by a priest, the brilliant Polish astronomer Copernicus, that the Earth and the other planets might more naturally be thought of as revolving around the Sun, this idea was considered more a curiosity, a handy way to calculate things such as planetary positions, but not the real *truth*. The Church insisted that the truth was that all bodies revolve around the Earth. Galileo's espousal of the heretical idea that the Copernican model was the truth eventually led to his arrest and confinement. The man with one of the greatest brains of his time was punished for believing the evidence of his eyes and the deductions of his mind.

Gradually other evidence added to the four Galilean moons and the world of science came to realize that Copernicus and Galileo had been right. Newton developed the concept of an attractive force called gravity that explained the motions of the planets and moons in marvelous mathematical detail. Believing evidence rather than doctrine became the rule when looking at the sky; this was a lasting contribution of the brilliant scientific pioneers of the seventeenth century.

Incredible Io

The innermost of Galileo's moons of Jupiter was named Io, after the hapless lass who was turned into a cow by Zeus. Io is similar in a couple of ways to Earth's Moon. It is just a little larger (its diameter is 2,200 miles) and it revolves around its planet in an orbit only a little larger than our Moon's (its distance from Jupiter is 250,000 miles). But, in most respects, Io is very different from our satellite. For instance, it revolves around Jupiter in only 1.77 days. As Newton would have explained, it *has* to revolve that fast in order to remain in its orbit around such a massive planet. Being close to a giant also leads to other remarkable features, the most important being the result of the tremendous tidal forces that push and pull on the moon, making it have the most violently active surface in the Solar System.

Although there had been some small evidence that Io might be a bit unusual before the Voyager mission, it wasn't until the pictures from Voyager 1 and 2 came back in 1979 that we realized what a bizarre world it was. Instead of a cold, bleak, cratered desert, its surface turned out to be brilliantly colored and covered with strangely shaped landforms, at first unrecognizable. When NASA published the first pictures of the face of this fantastic moon, the public claimed that it looked for all the world like a pepperoni pizza! Some of the features appeared to be something

Io as photographed by the Galileo spacecraft (NASA).

like volcanoes in appearance and soon it was realized that several of these features were, in fact, active volcanoes. Huge plumes of volcanic eruptions were sighted at the edge of Io and bright red and yellow lava flows were mapped on its surface.

The Voyager spacecraft detected nine active volcanic regions that produced giant plumes visible on the limb of Io. Comparisons of images taken at different times by the two Voyagers, which arrived at Jupiter four months apart, showed how the surface was constantly changing because of the ejected debris and flows of molten material. When the Galileo spacecraft arrived nearly 16 years later, it found that these same volcanic centers continued to be busy spewing forth their material and it detected many more places of activity, mapping a total of 50 active volcanoes. It not only witnessed plumes of volcanic material being thrown up visibly at the edge of Io, but it also was able to detect the glow of the volcanoes at night and in eclipse, when the surface was in darkness. Io glows in the dark!

A surprise with every orbit

One of the principal goals of the Galileo spacecraft was to study Io's volcanism in as much detail as possible. From June 1996 to December 1997 its ten close approaches to Io provided that opportunity, coming each time to within a few hundred thousand miles of the moon's surface. The planetary astronomers who followed its progress were not disappointed. Every orbit brought them new surprises.

Here is a highly abbreviated log of some of the new discoveries made with each orbit.

Encounter 1: new plumes seen, new hot spots recorded, plumes and hot spots seen to glow in the dark

Encounter 2: some plumes seen to have turned off, many surface changes noted since previous encounter, Pillar Patera darkens

Encounter 3: five large mountains newly recognized, layered terrain discovered on the side facing away from Jupiter

Encounter 4: new plumes detected, one nearly 300 miles high, new hot spots seen, changes seen in the surface near the north pole

Encounter 5: just coasting by, no data because of position of Sun directly in line with Io

Encounter 6: new deposits found around Ra in Encounter 1 seen to have faded away, many other surface changes mapped, the region directly below Jupiter seen to glow in the dark during eclipse

Encounter 7: new hot spots seen in several places, strange plume-like feature seen to glow in eclipse, the height of the prominent feature called Loki found to be negligible

Encounter 8: many new bright red spots seen to glow in the dark below Jupiter, surprisingly high temperatures measured for some hot spots, the glow of the atmosphere found to extend up to heights of 400 miles

Encounter 9: the hot spot at Pillan sending up a high new plume, very high temperatures measured, more new mountains detected, some showing tilted layering

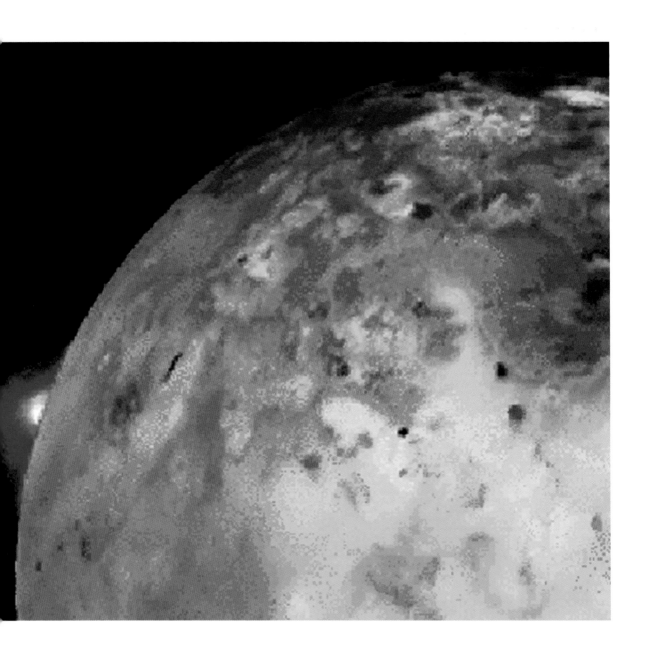

A volcanic eruption on Io in progress, seen beyond the left edge in this image (NASA).

Encounter 10: Pillan eruption leaves a huge circle of deposits on the surface, new active centers and new plumes detected.

This list of remarkable discoveries should give you a feeling for the tremendous level of activity on the surface of Io. Compared with our

Moon, where major changes occur on a time scale of billions of years, Io turns out to be able to pack in a tremendous amount of action in only a few months' time. The reason? Apparently it is largely Jupiter's fault. The tremendous tidal forces that the giant planet exerts on Io, amplified by the effects on its orbit caused by the next moons out, Europa and Ganymede, heat it up enough to make it a weird and wonderful world.

Monitoring Io's infrared radiation recently revealed just how much tidal heating there is. Io has a far greater total flow of heat outwards than the much larger Earth. It shines with the equivalent of one trillion (1,000,000,000,000) 100 watt light bulbs.

Is it lava or what?

One of Io's puzzles that astronomers hoped to solve with these encounters was the question of what that stuff is that Io's volcanoes spew forth. It doesn't look like normal volcanic rock like that found on the Earth. Its wild reddish and yellowish color is all wrong. The Voyager spacecraft did not answer this question satisfactorily and planetary scientists became divided in their opinions of what the material might be. Some were convinced that it must be made of sulfur and sulfur compounds. The bright colors could be explained that way, as pure sulfur can be brilliantly colored. If a chunk of sulfur is gradually heated, it starts out bright yellow, changes to orange, then becomes bright red, finally brown and then black. The pizza-coloring of Io has all of these colors and the presence of sulfur is a probable reason. Besides, astronomers detected large amounts of sulfur dioxide on the moon's surface, in its atmosphere and in the ionized cloud that occupies its orbit. It was therefore suggested that there might be a thick layer of sulfur and sulfurous material covering the surface of Io and that there might even be a sulfur ocean beneath the sulfur crust.

Other astronomers pointed out evidence that seemed to argue against the sulfur crust hypothesis. First, silicon, the common element of the other rocky bodies in the Solar System, is much more abundant than sulfur everywhere else. Why not believe that the "lava" of Io is made of silicates like the volcanic materials elsewhere? The mean density of the moon is just about right for a silicate body. Besides, close-ups of the huge volcanic craters on Io (it has many large calderas) show

As on Io, eruption of volcanoes on the Earth leads to widespread dispersion of ejected material. This is the volcano Vesuvius, which ejected ash that completely buried the city seen in the foreground, ancient Pompeii (author photo).

steep walls, and geologists point out that, if they were made of sulfur, these walls would collapse, as sulfur is not stiff enough to keep from avalanching. The temperatures measured by the Galileo spacecraft are too high for solid sulfur. Over the caldera called Pele, temperatures reach values as high as 700 °F, while sulfur melts at a temperature of about 250 °F. The steep, solid inner walls of Pele could not exist if they were sulfur.

The Galileo encounters with Io seemed to have settled this argument, at least to many scientists' satisfaction. It is probably silicate volcanism. There is definitely sulfurous material mixed in, but the bulk of the lava and ejecta must be made of silicates. This conclusion is, of course, indirect. We have not actually chemically analyzed any of Io's surface material. We have "seen" sulfur dioxide by detecting its spectrum in reflected light, but no spacecraft has sampled the material to determine without question the answer to the question, "is it lava or what?"

This chapter describes an expedition to the wild surface of Io for the purpose of answering this question directly. It will also provide a plan for exploring some of the remarkable landforms on Io, including a possible trip into an Ionian lava cave.

A safe landing

For a body as changeable and volatile as Io, choosing a safe place to land is not easy. It is especially difficult to do so for an expedition that lies very far in the future, as we do not know what surface changes may occur in the mean time. In this chapter we will suggest a landing place that may not be too hazardous, but you must keep in mind that it is primarily just an example and your team will need to use the latest surveys of Ionian activity to decide on a final landing place.

A goal of this trip to Io is to explore one of the active volcanic centers to learn first-hand about the nature of its colorful volcanism. There are many such centers and one must compromise between choosing safety and choosing lots of action. In this example, we will err a little towards choosing action, but you will want, of course, to consider safety as a first priority. During your orbiting period, you will be able to survey the surface in great detail and you can make a good choice then.

One of the most reliable volcanoes on Io is the one called Prometheus. Its high plume of volcanic material was seen throughout the Voyager missions and it was still active when the Galileo spacecraft examined it. There are other larger volcanic centers: for instance, Pele, which spews forth immense amounts of airborne material, making a huge bright red circle of deposits, but the scale of this prodigious object is almost too large for a reasonably short visit to explore. Prometheus, active and reliable, also has a relatively small source region which might be amenable to surface travel.

Famous for its reliable plume, Prometheus throws out material to heights of 50 to 100 miles above the surface. The deposits that this eruption leaves surround the source as a huge whitish ring. We believe that much of this stuff is a mixture of tiny particles and sulfur dioxide gas that is ejected when hot, molten silicates rise up from below the surface. There are other plumes that resemble those of Prometheus and scientists refer to them as "Prometheus-type" plumes. By contrast, the plume of Pele, prototype of "Pele-type" plumes, is larger, fainter and sporadic, leaving huge reddish rings of deposits. Besides these two kinds of eruptive plumes, there is a third type, called "Stealth" plumes because they are not visible by reflected sunlight. They are probably made up purely of gas, with no solid particles included.

Prometheus lies very close to the equator on the side of Io that faces away from Jupiter. Like our Moon, Io and the other Galilean moons rotate with the same period as they revolve around their planet, a result of the strong tidal action on them. Thus, when your lander puts down on the surface near Prometheus, you will not see the massive object that causes the violent activity around you.

Because of uncertainty about the nature of the ground on Io, it will be best to put down on it very carefully. It may be covered with a thick layer of dust that will fly up, obscuring your vision, or it may be just a thin crust that lies over a layer of sub-surface liquid. But the best images suggest that the surface is a nice, solid one, unusual in its chemistry, perhaps, but strong enough to land on.

The landing position that you choose will possibly be adjacent to the dark lava flow that issues forth from the caldera of Prometheus, flowing eastward in a meandering pathway out almost to the periphery of the circular ring of the Promethean ejecta blanket. There you can sample the volcanic products of this remarkable landform, collecting specimens of the lava and of the underlying surface materials for chemical and geological analysis.

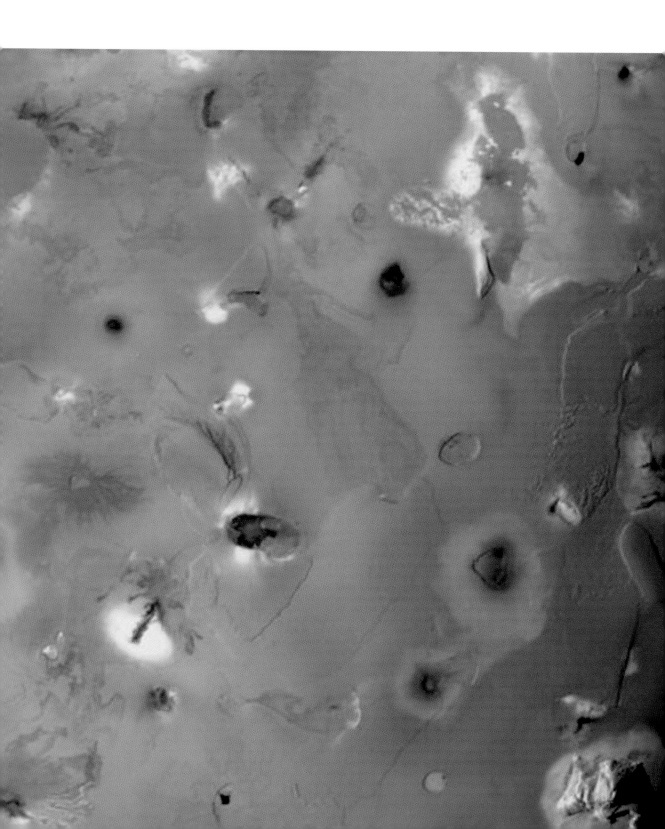

The large dark flow is probably mostly silicate lava, like basalt, but there is also probably some material in the area that is rich in sulfur and sulfur compounds, which you will want to study. Up close to the edge of the flow there is a surface rim that is bright white. Terrestrial observers have speculated that this is sulfur dioxide frost, formed when the hot lava melted surface sulfur materials, which later recondensed back as frost. East of the flow is a giant reddish area, probably a good place to examine for sulfur deposits.

[opposite] The south pole region of Io has both volcanoes and large non-volcanic mountains (NASA).

Prometheus' lava is probably made of silicates, like most lavas on Earth. However, there also may be pure sulfur lavas somewhere on Io. If there are any, they probably displayed some remarkable properties as they issued forth from their vents. When it is heated, sulfur goes through some unusual changes in its viscosity. Normal materials go from the solid phase to the liquid phase rather smoothly, with the liquid starting out as thick and viscous, but becoming more and more runny as it is heated. When solid sulfur is heated, however, it first becomes very runny and then as it gets hotter, it slows down and becomes thick and sticky. Only when it is hotter than about 300 °F does it again become less viscous. Were you to witness a sulfur eruption, the sequence of events would be amazing to watch. The flow would come swiftly out of the vent. Then, as it cooled, it would first become sluggish, slowing down as it moved down its valley. But upon reaching a certain cooler temperature, it would suddenly start flowing freely again, rushing ahead wildly before finally slowing again and eventually solidifying.

One possible adventure that you won't want to miss is the exploration of an Ionian lava tube. The probable basaltic flow from Prometheus may be similar enough to terrestrial flows that such tubes form there as the lava cools, letting the interior liquid lava flow out from a solidifying crust. On Earth such tunnels can extend for many miles, allowing people to walk or crawl through their slowly sloping, twisting length, a marvelous experience in exploration. To find such a cave on Io would be a special thrill and to penetrate into the interior of an exotic kind of lava flow might be an unparalleled adventure.

All of these features are spread out on a grand scale. The ring of Promethean deposits is 200 miles in diameter. In 1996, when it was mapped by Galileo, the lava flow snaked its way out from the center for a distance of 30 miles or so. You will need your Iomobile to cover all this ground. The days on Io will seem fairly long to you, as daylight lasts about

21 hours, with deep night lasting as long. You can split up into two teams so that you maintain a regular sleep schedule, each team alternately working outside in daylight and inside in the analysis lab. Your stay at Prometheus should occupy several Ionian days. You won't want to make it too short a stay, as there is much to learn and much to collect during this, your first Ionian adventure.

14

Mountain climbing in pizzaland

Half of Io's surface is a flat, barren desert – featureless plains of unknown origin. Perhaps these plains are vast lava flows that were smoothly laid down in the past or perhaps they were formed by years of the raining down of volcanic dust and cinders from nearby volcanoes. Whatever formed them was probably volcanic in nature and it must have happened recently. We are used to thinking in terms of billions of years when we ponder the lava plains on our Moon, but Io is very different. Its surface is continually being repaved, perhaps by as much as an inch in a year (terrestrial years, of course). So the Plains of Io, though smooth and blank, have got to be young – young, but probably dull.

The other half of Io's surface is anything but dull. It consists of weird and wonderful topography on a grand scale. There are two inhabitants of this varied land: volcanoes and massive non-volcanic mountains. In the previous chapter a plan was developed for visiting one of Io's hundreds of volcanic features, the colorful things that make it look like a pizza when seen from a distance. This chapter looks at the strange and mysterious Ionian mountains with an eye to making a spectacular first ascent of one of them.

Higher than Everest?

Can little Io, the size of our Moon, have mountains that are higher than the highest on Earth? Apparently so. There are hundreds of high mountains on Io and some of them rise up remarkably far above the flat plains. At least one of them, a weird and wonderful mountain whose name mostly consists of vowels, is the peak that we plan to send you up. Called Euboea, it is reported to be higher than Everest.

A listing of high mountains on Io that were mapped by the Galileo spacecraft in 1997–8 includes 95 peaks, but it is only a partial catalog. Of these, only 19 had names, having been detected by the Voyager mission years before and named by the official international commission that names these things. The rest are anonymous, though eventually they will have to have some kind of names. Unfortunately, it gets harder and harder to dig up names for planetary features, when so many are being discovered. So far the Ionian mountains have been given names from mythology, but there are only so many gods, sprites and elves. On Earth, we have had to duplicate names. There is York and New York, Cambridge, England, and Cambridge, Massachusetts, and Moscow, Russia, and Moscow, Idaho. Our mountains are rife with peaks called Bald Mountain, Green Mountain, the White Mountains and Mystic Mountain. Perhaps the Solar System will have to resort to some similar redundancy. In the mean time, we have a list of mythological names for Io, many of them rather awkward for English-speakers to know how to pronounce – Haemus, Euxine, Euboea and Boosaule, for example.

The heights of the mountains of Io can be measured in two different ways. One is to use the shadows the mountains make when the time is near dawn or dusk and the shadows are long. Knowing the angles accurately, scientists use simple geometry to calculate a mountain's height above the plains. Another method makes use of two photographs taken from slightly different positions in orbit above the surface, using the stereoscopic effect. If you look at two such photos with appropriate glasses, you can actually see these high, massive mountains rise up in front of you.

A long and detailed report on the Galileo mission to Jupiter was published in the journal *Icarus* in 1998. It included a list of 13 mountains for which accurate heights were known from shadow measurements. These ranged from a low 1,200 feet, not even half as high as the hills of England's Lake District, to 25,000 feet, high enough to rank among the high peaks of the Earth. But this was a very short list, as only a few mountains were caught making shadows at just the right time. Using stereoscopic views, planetary mappers discovered even higher peaks. For example, Boosaule apparently towers above its base to a height of 50,000 feet, almost twice as high as anything found on the Earth.

What made the mountains of Io?

There aren't many ways to make a mountain. On Earth, mountains tend to be built either by volcanic eruptions, by compression (where two sections of the crust push together, squeezing the land until it is pushed up, often in a jumbled mess) or by extension (where, for example, a section of land is broken apart and tilted). On Io, the mountains discussed in this chapter are not volcanoes. They do not have the form of volcanoes,

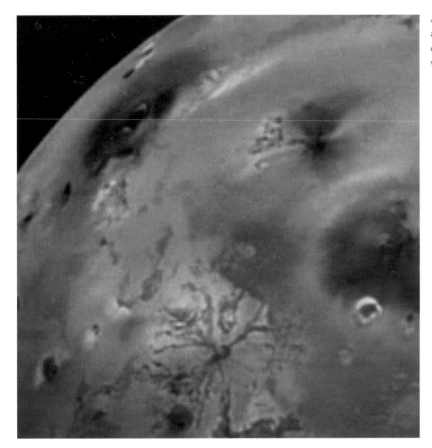

A Galileo image of volcanoes and calderas on Io. The reddish circle marks the ejecta from the volcano Pele (NASA).

nor are they usually associated with lava flows or craters. They are scattered nearly uniformly about the surface, not clustered near the volcanic centers.

Though they are not volcanoes, at least a few seem to have some connection with volcanic features. For example, there are some conspicuous mountains that rise up fairly near to the giant caldera named Zaal. And the famously active vent Pele has some tilted slabs of rock arrayed near it. The mountain Euxine, which rises to a height of 25,000 feet, has a large caldera near its south flanks and there is a small feature at its peak that might be a crater. But these are confusing clues. Though they suggest that volcanism may be involved somehow, the mountains themselves are not at all like volcanoes as we know them.

So how were they made? No one yet knows. Some are clear examples of huge tilted blocks of rock, looking like they were formed by giant forces

A map of the giant Ionian
mountain Haemus near
the south pole of the moon
(from a USGS map).

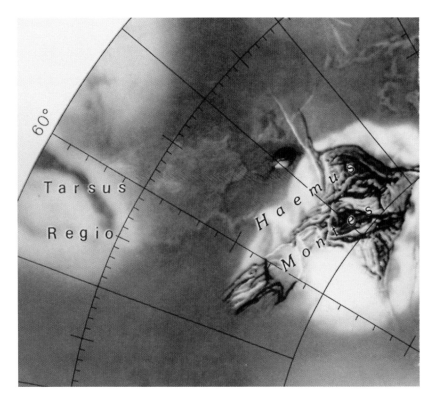

that broke off a section of crust and pushed it up to form a high blocky peak. The edges of some of these, such as Haemus, which rises to a height of about 30,000 feet, show parallel lines that might be layers of deposits, perhaps of volcanic flows. If so, they may be sections of the lava plains of long ago, now pushed up to dizzying heights. Possibly the mountains were formed by isolated faulting of the crust, followed by uplift and rotation. Geologists see a possible parallel with the mountains of the Pampas of Argentina or the Rockies of central Wyoming.

However, the global geology of these mountains is not like that of the typical mountain chains of the Earth. There are no long, parallel wrinkles like those that are so well-preserved in the Alleghenies and that make up most of the major ranges of the Earth. Instead each mountain massif seems to be nearly independent of others, each a chunk of crust that may have been broken off from the flat plains and pushed up from below. They do not seem to be the work of large-scale plate movement, but rather the work of smaller subterranian events, although no one knows yet what exactly these events are.

The architect's table mountain

One of the most spectacular mountains of Io is Mount Euboea, located in a fairly complicated area at middle-south latitudes. Shaped like a giant tilted table like those used by architects and artists, Mount Euboea presents so many interesting and typical features that we will choose it as the one of Io's mountains to explore and climb.

The mountain is very large. It is oval-shaped and extends over an area that measures 100 by 150 miles across. Compare this with Mt. Everest, which is only about 5 miles across. Its highest point is on the table edge at its southeast side, where it towers at least 35,000 feet above the Ionian plains, a mile higher than Everest. Southwest of it and almost touching it is a huge caldera with steep sides and a floor that is half covered with dark lava flows. To the north and west are tablelands, a few hundred to a thousand feet high, with edges that form steep cliffs like those of the mesas of the American Southwest. North of them is a major volcanic vent that has many black lava flows and white airborne deposits to proclaim conspicuously its volcanic nature.

Before considering how to climb this remarkable peak, it would be a good idea to examine it carefully to try to get at least a preliminary idea of how it was put together. First consider the steepest side, the high cliffs that form the southeast flank. These are spectacular cliffs, indeed, rising abruptly above the nearly flat southern plains. At their base there are two low steps, looking from orbit almost like giant examples of the preliminary steps you take before climbing the main steps of the library or, perhaps, the government building. Each is roughly a thousand feet high. From the top step it is about 20 miles along a fairly flat surface northwest to the actual base of the cliffs.

The cliffs themselves are huge and no doubt a close-up view will show many places where they rise almost vertically. However, the average slope is not steeper than the slopes on familiar terrestrial mountains. From its base to its crest, the angle averages about 40° or so. Perhaps an Earth analogy would be the Drakensbergs of southern Africa, which rise up steeply from the coastal plains of Natal and the Indian Ocean, though the Ionian cliffs total about three times the altitude difference. They are similar in the way in which the range is like a single tilted block of rock. From its eastern crest, the Drakensbergs slope gently westward slanting down to the colorful town of Maseru, capitol of Lesotho, 90 miles west of

A terrestrial mountain that shows conspicuous layering of ancient deposits. Subsequent to their deposition the land has been tilted similarly to what seems to have happened to Euboea (author photo).

the crest. However, the eastern cliffs of the Drakensbergs are sectioned by deep valleys formed by rivers, while those of Euboea are more continuous, there being no water erosion on Io. The surface of these slopes is hummocky and streaked with ledges and striations, perhaps revealing the layers of sediments or lava that made up the thick Ionian crust before this huge block was thrust up.

Turning now to the northwest flank of the mountain, we find a very different situation. It is smooth by comparison and its slope is a gradual 6°. Going southeast from base to summit it is nearly 75 miles. There is an abrupt rise all along the northwest base, somewhat reminiscent of the cliff at the base of Mt. Olympus on Mars. Above this the surface of the mountain flank is highly ridged. The ridges extend upslope as much as 30

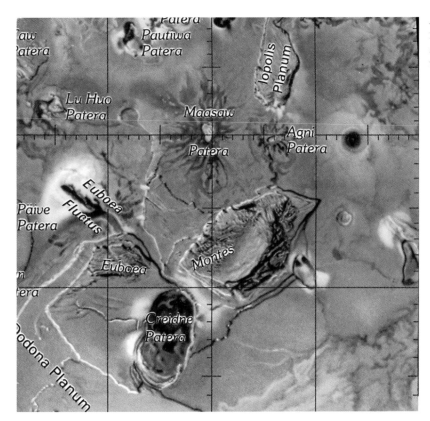

A map of the region of Euboea Montes. Various volcanic features are identified nearby (from a USGS map).

miles and are a mile or so across. They are thought to be the result of massive avalanches that brought debris down from above as the mountain grew. In character they look very similar to what geologists call "slope failures," the product of the uptilting of blocks of land that consists of a top layer that is weakly consolidated and that overlies a harder layer. There are no large blocks of rocks in the entire avalanche field, suggesting that the stuff that slipped was not all hard rock, but must have included some loosely arranged material, such as volcanic ash. The geologists who have studied Euboea liken these ridges to the giant avalanche formations on the Earth found near Carlson, Idaho, and to the avalanche slopes of the Ganges and Ophir Chasms on Mars (see Chapter 3). The difference is the immense scale of Euboea's avalanches. They have formed the largest avalanche debris apron that we know of in the entire Solar System.

The Drakensberg mountains of southern Africa are at the steep end of a massive uptilted region similar in scale and nature to Euboea (author photo).

Above these ridges the upper half of the broad northwestern flank is quite smooth. Except for some long arc-shaped wrinkles that conform to the shape of the curved mountain crest, spacecraft images show what appears to be a smooth sloping plateau. Apparently here there was little or no slippage and the surface is pretty much the undisturbed surface of this elevated chunk of the Ionian plains. From the point of view of a mountaineer, the upper half of this peak looks like the easiest part.

The climb

Higher than Everest, Euboea will be a major challenge to climb. The route is not so difficult. Were it on Earth, there would probably be no need for mountaineering skill or specialized equipment. It would be a long walk up, the upper part being so high as to require supplemental oxygen. But on Io there are all kinds of complications. One is that there is virtually no air to breathe at any altitude. The atmosphere of Io is extremely thin and patchy and it consists primarily of sulfur dioxide. So, as with the other adventures described in this book, you will need to wear a space suit to provide air to breathe, as well as a comfortable temperature.

Another complication that space engineers will need to have solved for you is caused by the strong Jovian magnetic field, the same that provided so much trouble when you descended into Jupiter's atmosphere. Even at Io's distance from Jupiter the danger from radiation could be severe.

The third complication is sheer distance. For Mt. Everest there is a certain amount of trudging through the Himalayan foothills before one reaches the bottom of the mountain, usually occupying just a few days. From there, the going is mostly up, a few thousand feet at a time. But for Euboea, the crest of the peak is 75 miles away from the base, meaning that you will need to spend an Earth week or so just to get there, regardless of any difficulties encountered along the way. It would be good to plan to have the Io Orbiter send down supplies at strategic locales so that you don't have to carry too much on your backs.

We plan the landing to be just to the north of the cliff that demarcates Euboea's northwest flank, at latitude 46°S, longitude 342°W. There is a prominent groove in the cliff just south of that landing place that probably would provide the easiest route up onto the lower slopes. The first climbing day will be occupied by this challenge, with the first rest period on the mountain being spent at the top of this valley. Resting will be somewhat awkward, as Io rotates with a period of about 41 hours, so that daylight lasts almost twice as long as your Earth-based body is likely to want to work for you without a sleep period. This might be a serious problem were it not for the fact that at night on this side of Io the landscape is nicely lit by the bright sunlit side of Jupiter. So you can plan to make progress at first in sunlight, then rest, and then wake up and proceed by Jupiter-light.

After the initial climb of the basal cliff, the climbing will likely be along valleys of rubble through the massive landslide debris that covers the lower slopes of the mountain. It will not seem very much like mountain climbing, as the upward slope is only a few degrees, an angle at which a comfortable road could be built (and maybe will be built some day). But it is not likely to be easy going, as there will be a jumble of boulders, rocks, dirt and dust to negotiate. It will take about three Earth days to get to the top of this rough terrain at an altitude of about 20,000 feet.

The last 15,000 feet will probably be easier, as the upper regions of Euboea appear to be fairly flat and smooth. Having followed the direction of the grooves that characterize the lower rough slopes, you will be headed directly towards the highest point on the mountain, which is near the center of the curved southeast ridge top. A good three days will probably be needed to reach this summit and you may want to spend another day there, drinking in the fantastic view and gathering geological samples.

What will the view be like? At night it will be particularly remarkable, as giant Jupiter will dominate the black sky and the landscape all around will be bathed in an eerie soft light, not unlike the moonlight on Earth during a full moon. Looking south from the rim, you will see in the distance the flat Ionian plains that extend off towards the horizon from the steep cliffs and plateaus directly below. To the southwest is the giant caldera called Creidne Patera, with its floor half-filled with black lava. Off to the north you may be able to glimpse eruptions of volcanoes. Perhaps Euboea Fluctus, a prominent volcanic vent to the northwest, will be active or another center of activity may have flowered in the time since the area was mapped by the Voyager and Galileo spacecraft. The small size of Io means that even at your great height the horizon will not be far away, so it will be difficult to make out such features as Maasaw Patera, a large caldera to the north, even though it has dark lava flows extending outwards for 50 miles in all directions.

In the Ionian daytime, Jupiter will still be there in the sky, but only as a crescent or gibbous "moon," depending on the angle with the Sun. The sunlight will sharpen the shadows that you see, making the surroundings appear more clear and stark. Surely the views, either at night or in daytime, will be a just reward for the strenuous and dangerous ascent to the top of one of the Solar System's highest mountains, higher than Everest.

15

Under the frozen sea

Jupiter's moon Io may be the most remarkable-looking moon in the Solar System, but the next moon out from it, bland Europa, may turn out to be the most exciting. As first seen from the Voyager spacecraft, Europa looked smooth and creamy, with gentle wisps of markings on its pale white surface. While Io appeared bizarre, Europa appeared soft and beautiful, like an impressionist painting painted by an artist who happened to be in a mellow mood.

The cosmic ball bearing

Close-up views of Europa, first from the Voyagers and later from Galileo, showed that its surface is almost perfectly smooth, like a giant ball bearing. There are no high mountains or deep canyons, just a few low ridges here and there and other minor wrinkles on a body that is the most remarkably perfect sphere in the Solar System. Europa is so smooth for a reason, but it took some time for astronomers to determine just what that reason is. Other clues about this unusual moon have emerged as exploration continues, so that now we have a pretty good idea why Europa has such a unique smoothness.

First, consider that you are looking down on the Earth from space. Where would you be likely to find the smoothest surface? The American Mid-west? The Steppes of Asia? The Great Sandy Desert? All of these features of the Earth are pretty smooth, but they are rougher than the surface of Europa. Instead, the places on Earth that most resemble this moon in its lack of elevations are the oceans. Even in a typhoon, the waves of the ocean are too small to ruin its innate smoothness.

This suggests that Europa is covered by water. But imagine now that the Earth is moved out to Jupiter's distance from the Sun. What would happen to our oceans? They would freeze, of course. This suggests that Europa may be covered with a frozen sea.

Another clue about this feature of Europa comes from its mass. Determined from its measured gravitational pull on the other Jovian satellites and on visiting space probes, the mass of Europa turns out to be 4.8×10^{19} metric tons (tonnes), a little more than half the mass of our

A Voyager image of Europa (NASA).

Moon. In diameter, Europa is a little less than 2,000 km, making it about 10% smaller than our Moon. Combining these measurements gives Europa's average density, which is 3.0 times the density of water, significantly less than the density of our Moon. If Europa is made up of rock like our Moon, then it must also contain some much lighter material. Water, one of the most abundant molecules in space, is a good candidate for this less dense material, so Europa's density also supports the idea of a global ocean. Making appropriate assumptions about the rock, it is possible to determine that there is a sea of water, at least some of it frozen, that is 50 to 100 miles thick.

There are more clues, one of which was discovered many years ago. The sunlight reflected from Europa's surface can be viewed with a spectrograph and the spectrum shows the unmistakable features of water or water ice.

Cracks and craters

The surface is smooth, but it is not liquid water. The vague, wispy markings that the early Voyager photos showed, on closer inspection, turned out to be permanent markings that were frozen in to the surface. The largest of these are giant cracks, some nearly global in length. Most are shaped like straight lines, leading to the scientists' name for them: the latin word *Linea*, but some are near-perfect curved circular arcs. When seen really close up, these cracks look like double parallel ridges, arranged almost like the two lanes of a freeway. Unlike freeways, however, when they intersect there is no cloverleaf interchange; one set simply crosses the other, obliterating it at the spot. This is how it is possible to figure out which set of ridges formed the more recently. But that is about

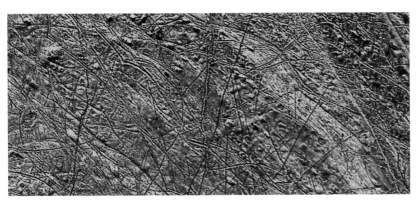

Intersecting cracks (lineae) on Europa's surface (NASA color-enhanced image).

all that astronomers have been able to figure out about these linear features, the origin of which is still something of a mystery. The youngest tend to be the narrowest and have the highest elevations above the plains. Galileo Orbiter images show them to be rugged, narrow ridges, a few really high ones rising up to about 1,000 feet above the plains. Between the two ridges is a deep valley a mile or so wide, the bottom of which is at plains level. Most, however, are very much more subdued, with multiple parallel lines sometimes spread out like a giant straight raceway, some 20 miles across or more. Their properties definitely suggest that they are in some way related to stress in the global frozen sea, some sort of cracks that filled with freezing water, but their remarkable detailed structure is still not satisfactorily explained.

Our first close-up views of Europa produced another mystery. Most bodies that have no atmosphere, like Europa, are covered with craters, formed by the collisions of the ages. Certainly meteorites, asteroids and comets are out there near Jupiter, as clearly shown by the heavily cratered surfaces of Jupiter's outer large satellites, Ganymede and Callisto. But the first Voyager images of Europa mysteriously showed no craters. The obvious answer to this mystery is that its surface is too young and fragile to retain craters after a body hits the surface. The frozen sea would soon fill up the hole with fresh water, which would quickly freeze.

This idea wasn't entirely abandoned when a few possible craters were found at last on the Voyager close-ups. But the Voyager's "craters," poorly defined and odd-looking, are not really craters, according to most scientists. They seemed to be fairly common, though nothing like as common as the craters on the Moon, and had led to the suggestion that the average age of Europa's surface is about one billion years. But we know now that this value is much too large. The Galileo images showed that these objects are probably not impact craters. Instead it showed a very few much clearer, undisputed examples. The real craters have unique properties that support the idea that Europa's surface is indeed quite young and must be replenished rather quickly after any outside disturbance. The very low numbers of these true craters indicate an average age for the surface of only 10 million years.

The craters show different characteristics according to their sizes. The small ones, less than 5 miles across, tend to look like fresh lunar craters, with crisp outlines and round bowl-shaped interiors. These small craters apparently have not broken into the depths of the crust very far and haven't had time for ice to fill them up. Larger craters, those 25 miles or so across, look a bit like large lunar craters with central peaks and walled rims, but they are much flatter than lunar examples, probably because the icy surface is too plastic to hold any great range in elevation. They seem to have "relaxed" after the crash, flattening out, but maintaining their outlines. A particularly nice example is the crater named Pwyll, about 15 miles in diameter, with a beautiful range of "hills" near its center and a well-defined "rim," but with hardly any variation in elevation from outside in. It has a set of bright rays radiating out from it, probably ejected material thrown out on impact, as in the case of Copernicus on our Moon.

Even larger craters seem to exist, but they are not like any craters seen on the Moon. A good example is the giant feature named Tyre, which is a near-circular bull's eye about 75 miles across. It consists of an immense set of concentric rings, made up of dark bands and low ridges. At its center is a smooth, bright area, about 10 miles across. If Tyre was formed by an impact, the asteroid or comet that caused it was large enough to penetrate completely through the frozen skin of Europa's putative ocean, causing a huge set of circular ripples to form in the ice around it. The central smooth place may be where the collision destroyed the ice, and water from below replaced it, forming a fresh, bright new icy plain. There is quite a story told in the ice of Tyre: it lies on top of several older linear cracks, but its western edge is cut by two wide ridge pairs, formed after Tyre was formed. There are several other, more narrow, cracks that cross

A Galileo image of the impact feature Pwyll on Europa (NASA).

the bull's eye, indicating further tectonic activity. It is not known whether these newer linear features were formed right after the impact or not; perhaps some, at least, show us how the impact might have disturbed the crust of the moon enough to break it up into these mysterious double-ridged lines in the ice.

Changing diapirs

In some areas on the amazing surface of Europa, for example at a place named Conamara Chaos, there are icebergs, great chunks of icy surface that seem to be floating in a frozen sea, much like we see in the Earth's polar oceans. These pieces of icy surface are typically about 5 to 10 miles across with about the same separation and they appear to be displaced from their original positions by crustal movement. They are tentatively identified as what geologists call "diapirs," which are buoyant chunks of material that are created by instabilities that result from differences in temperature. It may be that the various dark reddish spots, pits, and moats that are found on Europa are the sign of convective movement of ice in the outer layers above the sea. The darkish areas may be pieces of ice from below, where it is warmer, brought to the colder surface by diapiric convection. The Conamara Chaos region may be a concentration of such activity. It certainly looks chaotic, like certain areas of the frozen arctic sea of the Earth in winter, when channeled ice islands apparently float in a solid sea.

Areas like Conamara are young and seem to have been changing constantly as chunks of material rise to the surface and others sink to the

An area of giant iceblocks on Europa (NASA).

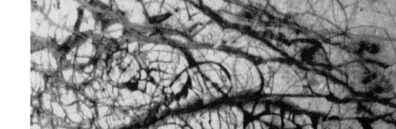

depths. We can't now predict what these areas of chaos will be like when your expedition arrives. It will be an important task for you to photograph them from orbit so that geologists back on Earth can compare their present condition with that when the Galileo images were obtained in 1998–9. Look at them carefully; their properties suggest that they came from below, from the region of the ocean that is your main goal to explore, so their color, their chemistry and their physical characteristics may be important clues to the mystery that you aim to solve.

Ice climbing

It's tempting to plan an expedition to the amazing chaotic terrain of Conamara. The jumble there of giant icebergs would be a spectacular place to explore and these huge chunks of European crust might be useful as clues to the nature of the sea that lies below. If your expedition is relaxed enough to make two landings on the surface, then this would be a tremendously interesting place for the first one. Spending a few days among the icebergs and rough ice between them looking for ways to infer the geology of the layers below may well be scientifically profitable. Certainly it would be an adventure. Even finding a smooth landing spot might be tricky. Moving about on the surface would be a mountaineering-style challenge. Climbing the ice cliffs and avoiding the likely crevasses will probably be quite a bit like ice climbing on glaciers on Earth's mountains.

Ice fishing

But the main goal of this trip to Europa lies below and Conamara is probably too disturbed a place to choose for it. Instead, you will travel to the center of Tyre, the giant impact feature that has a nicely smooth central area that would be an ideal landing spot. This flat circular center is probably the impact point of Tyre, where the asteroid or comet penetrated the crust, and the explosion of the impact may have created an open shaft that reached down into the ocean below. If so, then the smooth circle lies on top of a vertical tunnel of fresh oceanic ice. Your crew will want to study the surface ice carefully for what it can tell you about the chemistry of the sea. They may find evidence of other, more exciting constituents of the ice, too.

Icebergs on Earth. These were photographed in August in Scoresbysund, Greenland (G. Wallerstein photo).

We have chosen the center of Tyre for a reason similar to the argument that led scientists to propose that prospectors in Sweden drill down below the surface of one of its large meteorite impact craters. The feature is called Siljan, named after one of several scenic lakes that nearly fill the structure, which is about 30 miles across. The idea at Siljan is that the impact probably penetrated the crust of the Earth and therefore drilling down through this break in the solid Swedish granite might provide easy access to a wealth of natural gasses, thought to be trapped far beneath the surface. Although the natural gas has not so far been found by drilling, the idea is a good one to adopt for Europa. The impact at Tyre probably also punctured the crust, providing a clean route to the goal of your ambitious expedition.

At last report the drilling at Siljan had reached a depth of about 20,000 feet. Coincidentally, this is probably about how deep your drilling team will have to go to reach the bottom of Europa's icy crust. It may be less, though, as we hope that the impact may have left a channel below that has not yet fully frozen. Of course, we don't actually know at this time how deep the crust is; a depth of 5 or 6 miles has been inferred from various clues at the surface. By the time you are there, however, this fact will have been established more securely.

Drilling to such depths is not an easy task anywhere and it certainly will take a considerable time on such a foreign and exotic land. It may be a month or so before the drilling crew will be able to announce that they

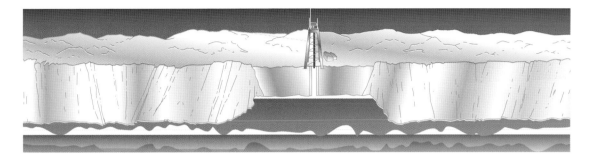

have finally reached the bottom of the ice. During that interval, your team of explorers will want to travel across the icy plains of Tyre, exploring the sequence of circular ridges and the criss-crossing and intersecting lines, so as to build up a geologic history of this complicated place.

On the icy surface of Europa a well is being drilled through the shallow ice of an impact crater to provide access to the mysterious ocean beneath.

Living for a few weeks on Europa will be a memorable experience. Of course, it will be very cold. Daytime temperatures will be about $-200\,°$F. The Sun may warm you a little, but it is five times farther away than it is from the Earth, so its light and heat is 25 times less, making a clear day on Europa about as dark as a very cloudy day on Earth. Daylight will last somewhat less that two Earth days (about 43 of our hours). The surface gravity on Europa is less than on Earth and even a little less than on our Moon, so getting around with your heavy space suit will not be too awkward.

Meanwhile, back at the well, the drilling will have finally reached the briny deep. We believe that the European ocean is, in fact, briny, because the satellite has a magnetic field that is probably caused by currents in the ocean, which therefore must be salty or else it would not set up a magnetic field. Chemical tests will be made on the water that gushes up from below, of course, but most of the crew will be more interested in another aspect of this foreign water. Does it show any signs of living organisms? The fact that it is probably salty together with its warmth have led scientists to speculate that life may have evolved in this subsurface ocean. We have no idea what sorts of life might be there, but the possibility of at least microscopic biology is an intriguing thought. The discovery of flourishing life at the bottom of Earth's oceans, where it is warmed not by the Sun but by volcanic fumeroles, has encouraged biologists to think that life might exist in exotic places, as long as water, minerals and warmth are there. Of course, the life forms are unlikely to be very much like us.

Descent into darkness

If there is no sign of life in the water, it may be that you will decide to gather your data and your samples and return to Earth. However, if you do see even weak evidence that Europa's ocean has life in it, your plan will be to explore that dark and unknown world in a more direct way. We have plenty of evidence that on Earth, if there is microscopic life, then there may well be macroscopic life, as well. Possibly the micro-organisms that your biologists have discovered in the water are food for predators, life forms too large to have been brought up by the drilling. You should, in that case, go down there to find them.

Drilling a narrow well to penetrate down through the ice will have seemed an almost impossible task in this remote site, but it will look simple compared with the formidable task of enlarging the shaft to be big enough to allow descent of your exploring submarine. Nevertheless, the task must be accomplished, even if it takes many weeks. And the shaft must be kept warm and open during the whole process, so that the terrible cold at the surface does not freeze up the water before you can complete your trip.

Your submarine will probably be lowered by cable, the 5 or so miles of cable being made of superstrength fiber specially developed for this purpose. However, you will probably want to be freed (temporarily) from the cable when you reach the ocean depth so as to be able to explore more freely.

What will you find down there? We really have no idea. It will be a true adventure, exploring a pristine world. There may be huge, weird ice formations, resembling the fairyland features found in terrestrial caves. And, of course, there may be huge, weird life forms. You will hope that these, if they exist, will not turn out to be so much bigger than the submarine that they think of it as exotic food. Will these possible Europeans have intelligence like our whales, able to recognize you as a friendly visitor? Or will they simply look upon you as a Jovian Jonah and swallow you without asking questions? As in any major adventure, there may be some danger in this exploratory journey, but the rewards of the knowledge gained and the thrills of discovery will be there, too, deep in the Solar System's darkest and most mysterious ocean.

16

Snowboarding through Saturn's rings

When we were very young, we took great pleasure in bumping into things. Of course, when we were infants learning to walk, bumping into things came naturally and that may be what gave us a taste for that sort of activity. The pinnacle of happiness came when we were old enough to pester our parents into taking us to an amusement park, where we were uncontrollably attracted to the bumper-car arena, and, with screams of delight, we steered our heavily bumpered little electrical cars into every other heavily bumpered little electrical car, whether or not we knew the other drivers, friend, foe or stranger. If this is one of your childhood memories, the adventure described here may appeal to you, in spite of (or perhaps because of) the rather terrifying danger involved.

Everyone's favorite planet

If you should ask any assembled group to name its favorite planet, the winner will surely be Saturn. From the first grade class to the nursing home, almost everyone thinks of Saturn as the beautiful one, the one most likely to stir feelings of awe. When observed through a telescope on a good, steady night, Saturn is a wondrous sight, likely to cause a sponta-neous reaction of "Oh, wow!" or "Cool!". The planet itself, however, is really pretty bland. Only really sharp eyes can make out any details on its cloudy surface: faint, wispy bands that seem to come and go. It's not the planet that wows the observer, it's the rings. Bright, symmetrical, set at a jaunty angle, Saturn's rings are the Solar System's best. The other three giant planets also have rings, but they are nothing like Saturn's, being barely detectable from the Earth, even from space telescopes. Saturn's are one of the wonders of the planetary world, well worth a special exploratory expedition.

Ears or what?

When Saturn was first observed with a telescope by Galileo, he was mystified by its strange appearance. He and other seventeenth-century astronomers often drew it as a triple planet, a large one attended by two

A Voyager image of Saturn and its rings (NASA).

smaller ones on either side. In some cases they were drawn as giant Saturnian ears, turned our way as if trying to hear our exclamations of amazement. As telescopes improved, better clarity showed that the aspect of these strange things at the sides of Saturn changed over time. Sometimes they seemed large and then, months or years later, they almost disappeared. More than 50 years passed before telescopes were good enough to show them as what they are: thin rings that extend out from above Saturn's equator to dizzying heights above the planet.

The outermost edge of the bright rings lies more than twice as far from Saturn's center as its surface. Because Saturn is a very large planet, with a diameter almost ten times the Earth's, this is a large distance. The ring's edge is about 82,000 miles out. The inside of the main rings is about 55,000 miles from the center of Saturn, meaning that it hangs 20,000 miles above the planet's surface. Although we are not planning this expedition to include a visit to Saturn's surface, it would make a nice detour just for the spectacular view you would have of the sky. The rings would be dazzling sheets of light arcing across the dark Saturnian sky. Their brilliance would be most impressive just before dawn or after dusk, when the Sun would be below the horizon, but the illuminated parts of the rings would be spectacular curtains hanging above you in the star-filled sky. Note, however, that you would not be standing on Saturn's surface to see this view. Like Jupiter's, the surface is a deck of clouds that extends hundreds of miles down into the planet's interior, which is mostly made up of liquid molecular and metallic hydrogen. There is nothing to stand on.

Gaps of mystery

People speak about the "rings" of Saturn, but why not just the "ring"? Its plural nature was first realized in 1675 by Gian Domenico Cassini, a talented Italian astronomer who had been lured by Louis XIV to France to direct the Paris Observatory. The excellent telescope there and his perceptive eye allowed him to see a thin line that divided the ring into two nearly equal halves. This gap in the ring structure is referred to as "Cassini's Division" and it is a good test of the quality of a small telescope. Modern measurements show that the gap is about 3,000 miles wide.

A second, smaller gap was detected somewhat later by Johann Franz Encke. Encke's Division is only about a tenth the width of Cassini's and is difficult to see from the Earth except with a moderately large telescope and excellent atmospheric conditions. It lies about 2,000 miles from the outer edge of the outer bright ring.

The discovery of gaps led to the necessity of having names assigned to the different rings. Cassini's Division separates what astronomers

Backlit view of the rings, like a view that might be had from the surface of the planet (NASA).

came to call the "A" ring, the outer portion, and the "B" ring lying inside the gap. In time another ring, much fainter and more transparent, was identified and called the "C" ring. It extends inside the B ring and can be traced down some 8,000 miles towards the planet's surface.

The gaps in Saturn's rings remained a mystery for two centuries. No one had any idea what the physical properties of the rings were and no physical principles were applied to them to attempt to understand their amazing appearance. History suggests that astronomers in those days were content to describe an object in the heavens without questioning the "why's" of its properties, probably because physics was a young and slowly growing discipline. It is hard now for us to imagine anyone being happy with a description of a mysterious natural phenomenon without trying to understand why it is like it is.

But no one could have understood the gaps until there was some clear idea of the physical nature of the rings themselves. Were the rings giant sheets of some solid substance that formed in some remarkable way to hang there like the rim of a hat? For two centuries this question was debated by astronomers; Cassini himself, as well as his son, who succeeded him at the Paris Observatory, claimed that they must be instead made up of swarms of little moons.

By the middle of the nineteenth century, astronomers were finally able to settle the question. In 1859 the talented mathematical physicist James Clerk Maxwell showed that the rings could not possibly be solid. He proved mathematically that solid rings would be completely unstable and would break up into pieces that would then orbit the planet like tiny moons. A direct demonstration of the truth of this conclusion came several decades later when James Keeler used the largest telescope in the world at that time, in California, to measure the velocities of the rings. He found that the outer parts of the rings revolved more slowly than the inner parts, obeying exactly Newton's laws for orbiting satellites. He showed that the rings do not revolve like a phonograph record, but behave like thousands (actually, billions) of small orbiting moons, just as predicted by Maxwell's physics. Actually, anyone can prove for himself or herself that the rings are not some solid material, like stainless steel, by observing Saturn when it passes in front of a star. The star remains visible when it is behind the rings. They are transparent, as the particles are loosely packed, allowing light to pass through.

Mimas, an inner satellite of Saturn (NASA).

So what causes Cassini's gap? This question was answered when astronomers noticed that Cassini's Division was at the exact place in the rings where the moonlets have a period of revolution that is just half that of the moon Mimas. The innermost of Saturn's larger moons, Mimas, was discovered in 1789 by Sir William Herschel. Its period around Saturn is just under one Earth day, meaning that it really barrels along, to cover its 700,000 mile orbit in such a short time (this rapid orbital speed is the result of Saturn's large mass; if Mimas moved more slowly, Saturn's gravity would cause it to plummet down into the planet). The period of ring particles that would be in Cassini's Division is just under half a day. Astronomers of the nineteenth century realized that this would set up what is called a resonance, so that the gravitational pull on a particle by Mimas would build up to the point that the particle would move to a non-resonant orbit. In that way the gap is cleared and maintained.

Thousands of rings

When the first Voyager spacecraft arrived at Saturn back in 1979, it revealed some additional rings, named (unimaginatively, but following tradition) the D, E, F, and G rings. The D and E rings are very faint and diffuse, barely visible above the sky background. The D ring is really just an inner extension of the C ring, ranging from its inner edge all the way down to the cloudy surface of the planet. From a spaceship orbiting close to the cloud deck, you would probably be able just to make it out as a tenuous sheet of ghostly light, extending up from the equator towards the brilliant rings above you.

The E ring, on the other hand, is a very different affair. Its faint light was first detected from the Earth, but it was the Voyager images that allowed it to be mapped and named. It lies outside the main system of rings, with its inner diffuse edge about 25,000 miles above the outer edge of the A ring, and it extends 180,000 miles farther up above the planet. Remarkably, this brings it out amongst the inner main moons of Saturn. The moons Mimas, Enceladus, Tethys and Dione all plow right through the E ring as they spin around the planet. When it was noticed that the E ring is brightest right near the orbit of Enceladus, astronomers guessed that it might be derived from that peculiar moon as debris ejected somehow from its surface. Voyager images of Enceladus showed it to be much brighter than the other moons near it, suggesting that it must have a fresh icy surface. Perhaps it suffers volcanic-like eruptions, triggered by tides as in the case of Jupiter's Io, and these eruptions eject pieces of ice into space, where they go into orbit around Saturn, forming the E ring. Enceladus may be the Solar System's champion snowball thrower!

The F and G rings are faint and narrow. They both lie outside the bright rings, but inside the orbits of the main satellites. When first seen by Voyager they presented a major mystery, as they were so narrow and crisp. Astronomers asked, "what keeps them from dispersing and becoming diffuse like the other rings?" The answer soon came when it was discovered that these rings act like sheep. Two small moons, named Pandora and Prometheus, orbit on either side of the F ring. Called "shepherd satellites," these moons act as gravitational shepherds; when a moonlet in the ring strays outside its narrow confines, the shepherd satellite comes along and interacts gravitationally so that the moonlet goes back into the ring. The inner shepherd, Prometheus, moves faster than the ring's moonlets, so that a wanderer that moves into a smaller orbit is soon encountered by the faster Prometheus. As it approaches, its gravitational pull slows down the moonlet, causing it to fall back out into the ring. Pandora does a similar job to keep the moonlets from wandering beyond the outer edge of the ring. The fainter G ring lies even farther out and seems also to have its shepherds.

So we have rings A through G. This is only seven rings, but this section is entitled "thousands of rings." How can that be? The answer comes from another remarkable discovery made by the Voyager spacecraft. Just as you can see stars through the rings, astronomers were able to trace the spacecraft when it passed behind the rings. They recorded the signals as it did so and discovered rapid fluctuations, indicating that the ring particles are actually divided into thousands of little rings all stacked together. These are the result of various gravitational forces, caused by Saturn's moons and the moonlets themselves. The best images of the rings from the Voyagers show the most conspicuous of the ringlets that comprise the main rings of Saturn.

Snowballs and icebergs

OK, that's enough of all these preliminaries, what's the plan? This adventure is quite different from those of earlier chapters. In Saturn's case, we are not climbing a mountain or making a trek through some wilderness landscape. We'll instead be hurtling through one of the most hazardous zones in the Solar System. If we want to survive, we'll need some special knowledge and some extra skills. Our plan is to plow straight through the thickest of Saturn's rings and back again, bumping our way among thousands of hurtling fragments.

Fragments of what? What are these things that make up the rings? Giant rocks? Tiny pieces of dust? Actually, they are neither of these. A clue to their nature is the fact that they are so bright. They reflect sunlight very well, as if they were something that is nearly perfectly white. Giant rocks or pieces of dust would not look like that. Early on astronomers guessed that the rings might be made up of water ice, which would be that white. A few decades ago this surmise was confirmed when telescopic evidence showed that the rings reflected sunlight exactly as water ice does, leaving its fingerprint in the infrared part of the rings' spectrum.

So the rings are ice, but are they giant chunks of ice like glacier-covered mountains or are they a fine mist of ice like terrestrial cirrus clouds? The answer is important to know. Were we to fly a plane through one or the other on the Earth, the result could be very different! Radar studies of the rings from the ground helped to explore this question, but it took the Voyager visits to answer it definitively. By studying how the sunlight is reflected both backward and forward (yes, light can be reflected forward; it is called "forward scattering" and is a characteristic of small particles), astronomers were able to learn the sizes of the ring fragments. Most turn out to be the sizes of snowballs and hailstones, with some that are big enough to make a snowman if a couple are stuck together and a few that are as big as igloos or even modest icebergs. There is also a great deal of snow – fine particles of ice, the kind that leave a fine dusting of snow on your car after a dry snowstorm on Earth. These small particles are especially abundant in the B ring, which is the densest and brightest ring.

We don't really know what these snowballs and icebergs look like. They might be round and smooth, the result of constantly bumping into each other. Imagine a freeway of reckless drivers who are driving cars made of aluminum foil. Regardless of how fancy and fashionable these cars might have looked back in Detroit, it wouldn't be long before constant collisions would turn them all into wrinkled spheres, ready for the recycle bin.

Most scientists suppose, however, that the ring fragments will turn out to be quite irregular in shape. Perhaps the largest, the ones as big as icebergs, will not be solid ice, but rather will be loosely packed piles of ice rubble, maybe a little like the ammunition dump in a neighborhood snowball war. As in a particularly nasty war, some of the snowballs might have rocks in their centers, a possibility to remember as you navigate through this hazardous hailstorm.

These issues might seem to cloud the enterprise. When there are so many uncertainties and so many possible dangers, it is sometimes forgotten that an exploration can be both a beautiful experience and lots of fun. Up close the rings are probably a wondrous sight. From orbit just above them, they will be spread out below like a glorious, immense snowfield, with the thousands of ringlets nestled inside each other and extending like bright, icy lines off into the distant horizon. And above them to one side the giant dome of Saturn will dominate the dark, starry sky. It will surely be a breathtaking sight, worth savoring before descending into the fray.

Into the blizzard

The plan: your spacecraft will orbit Saturn just above the B ring at a distance of 30,000 miles from the surface of the planet. The captain will need to apply power to the craft very slightly to prevent it from crashing into the ring, as a natural orbit will, of course, have the center of Saturn as its center. Without power, the spacecraft would spend half of an orbit above the rings and half below during each orbit, passing through them twice, a possibly dangerous move. The orbital speed will be about 60,000 miles per hour and the spacecraft will orbit the planet once every 11 hours. Days and nights will flash by about twice as rapidly as on Earth, so you'll probably adjust fairly well to the schedule, awake and asleep during one day and one night each.

Your spacecraft's speed is tremendous, of course, as it must be to prevent its falling into such a massive planet. But looking down at the rings, you will not think that it seems so swift, as the rings will be revolving at the same speed as you (provided, of course, that your captain hasn't forgotten his instructions and put the spacecraft into a retrograde orbit, in which case the ring particles would be traveling 120,000 miles per hour in the opposite direction!). The key to your safety is the fact that you need to be hurtling through space exactly together with the snowballs and thus any collisions (and there will be many) can be gentle ones that will not fatally destroy your space suit.

The title of this chapter, "Snowboarding through Saturn's rings," suggests that you might be using some kind of ski or board-like apparatus to make your trip more fun. In fact, your Saturnian snowboard is only slightly like those used on terrestrial snow slopes. It looks quite different and it has a fairly different purpose. A snowboard confines your feet, allowing you to pummel down a slope without being able to stick one foot out to prevent a fall, thus encouraging you to make a spectacular collision either with the snowy ground or with another snowboarder. The Saturn model, on the other hand, is designed to prevent collisions. It acts more like a bumper that lets you plow through the ice and snow, leaving your space suit intact. While the terrestrial snowboard mounts on your feet alone, the Saturn snowboard fits over most of you, like a giant cocoon, with openings for your eyes and your hands and feet, but with protection for your body in general. It is like the famous Whipple Bumpers that were designed years ago by the astronomer and space pioneer Fred Whipple, who realized that a spacecraft would need an outer skin to protect it from interplanetary dust and meteors. In your case, of course, the danger of bumping into something is tremendously enhanced there in the thick of Saturn's rings.

After your spacecraft has orbited Saturn a couple of times, your scientific colleagues will have carried out enough studies of the rings below you to be able to say just how close our predictions were to the true character of the ring fragments. Perhaps you will need to make adjustments to your plan of exploration. But there will probably be no need to drastically change the plan. Soon you will be ready to don your snowboard and step out through the airlock chamber and thence to the rings.

The step from spacecraft to rings is a big one. To be sure that the spacecraft is safe from collisions with icebergs, it will have maintained a distance of some 10 miles above the top of the rings. The rings are actually very thin, only about 30 feet thick, but they are slightly bent, like corrugated cardboard, with the bends ranging over a mile or so. Your first giant step will be a leap of about 10 miles, which you will want to make fairly rapidly, but under power so that you can stop when you reach the ring surface. Plan to take a couple of hours to make the descent and then, when you are at the fringe of the rings, plan to spend another hour photographing and testing just out of range of the missiles.

About three hours into the mission you will be ready for the main event. Slowly, you will descend into the blizzard of snow and icy projectiles. As you bounce about, snowballs will be deflected gently from your snowbumper. You will be able to avoid crashing into the rare house-sized icebergs by grabbing a snowman-sized chunk and pushing off in a chosen direction. The fun of careening slowly among your icy orbiting companions in space will be a dramatic and thrilling demonstration of Newton's laws. Of course, in spite of the fun of the gravitational games you can play, you must remember to take plenty of photographs and videos, and also to gather a number of samples of the smaller snowballs to pack back to home.

The rings are remarkably thin, and it won't take you long to traverse their entire thickness. They are only about as thick as a large living room, so you can expect to bump your way to the bottom of the rings in only a few minutes. The remarkable thinness of the rings results from Saturn's great gravitational pull on them, which restricts them to the equatorial plane of the planet. When you pop out on the other side you will probably notice how symmetrical the rings are; they look just the same from either side.

Back inside the rings, you can continue your exploration and sample gathering for another hour or so and then it will be time to ascend back to the spacecraft, which has been hovering above you, listening to your verbal descriptions and monitoring your progress. Using your small rocket engines, you'll power yourself back up slowly, to be welcomed back by your excited (and probably envious) colleagues.

Saturn's rings are among the most spectacular objects in the Solar System, but only you will know from first-hand experience that they are

What it might be like to be among the icy missiles in the thick of Saturn's rings.

also among the flimsiest. It may seem ironic that you will have traveled some 800 million miles from Earth to traverse a snowfield that's only 30 feet deep. But its amazing thinness and the remarkable sort of bouncing about that you will have experienced are both spectacular examples of the power of the law of gravity, a law that all of us must obey and therefore one that we should thoroughly understand.

17

Titan's tarry seas

Looking at Saturn through a telescope on Earth will not only show its glorious rings, but even a small telescope will also show its largest moon, Titan. Seen as only a small yellowish dot from here, Titan is actually a giant moon as seen from up close. It is larger than the planet Mercury and has many features that resemble those of the Earth. The most Earth-like is the fact that Titan has a nice, thick atmosphere, made up mostly of nitrogen, as is the Earth's. Unlike the puny atmosphere of Mars, Titan's is thick enough that the pressure at its surface is very much like ours; in fact, it's about 50% greater than the air pressure here. Where we had problems with the scarcity of air on Mars and with the overabundance of it on Venus, Titan has just about the right amount to make us feel at home.

But Titan is not really very homey. While it's true that its atmosphere's main kind of gas is the same as ours and the pressure is like ours, the temperature is something else. Although we've not yet seen the surface, we know that it is not a hospitable place, as its temperature measures about $-300\,°F$! The adventure described in this chapter is a really difficult one to plan. Titan is horribly cold and the nature of its surface is still shrouded in mystery. Reaching and exploring that surface will be a tremendous challenge.

Discovering an exotic new world

After Galileo's amazing discovery of Jupiter's four large moons, people began to realize that the planets were worlds like Earth. They could have satellites that travel around them, a revolutionary idea at a time when doctrine had taught that everything in the universe revolves around the Earth.

Another exciting discovery was made 45 years after Galileo turned his telescope toward Jupiter, when the Dutch astronomer Huygens focused his long, unwieldy telescope on Saturn and realized that a faint star-like object to one side of it was a moon. It was later named after the classical god Titan, from the same mythology that gave Saturn its name. From watching Titan's slow motion in the sky, it was clear to Huygens that it orbits Saturn. Its period is just about half that of our Moon, about 16 Earth days. If Saturn's calendar is governed by its main moon, Titan, then

Saturn's months are half as long as ours, but its year, of course, is much longer, 30 times longer than ours.

Saturn has many other moons, but none is as large or as intriguing as Titan. In the 20 or so years following Titan's discovery, the astronomer Cassini, who discovered the Cassini Division in the rings, found four more moons of Saturn, all of them quite a bit fainter than Titan. With a diameter of a little over 3,000 miles, Titan dwarfs these others, none of which is larger than a thousand miles. None of them has an atmosphere, either, as all are too small, with too weak a gravity to hold onto enough gasses to keep a permanent atmosphere.

Many other small moons orbit Saturn. Including the little ones that shepherd the rings, Saturn's family consists of 18 moons with names, plus a dozen newly discovered unnamed moons and the millions of moonlets that make up the rings. Viewing the sky from orbit above Saturn's clouds, astronauts will need to consult a schedule to figure out just which moon is which up there above them in the ring-filled sky.

Three hundred years after the discovery of Titan, the astronomer Kuiper made an equally-exciting discovery. From his telescopic study of the spectrum of the light from Titan, he found the telltale signs of a gaseous atmosphere. The gas was methane, which is made up of hydrogen and carbon atoms. Methane is also found in our atmosphere as a rare (and dangerous) constituent, but is more abundant under ground. It makes up part of the natural gas that fires many of our stoves and furnaces. It is also a gas that can exist in gaseous form even at very cold temperatures, which explains why this same gas has also been detected in the atmospheres of Jupiter and Saturn.

Kuiper's discovery made astronomical headlines. It was the first detection of an atmosphere on a moon. The fact that the gas was methane interested even biologists, who came to realize that Titan's atmosphere might provide them with important clues about how the Earth's atmosphere evolved, back in the days when life was first forming on our planet. In Titan's case, however, it was eventually learned that methane is only a minor component.

Smog alert

Before the Voyager spacecraft reached the Saturn system, we knew very little about its moons. Titan was known to be very large and to have an

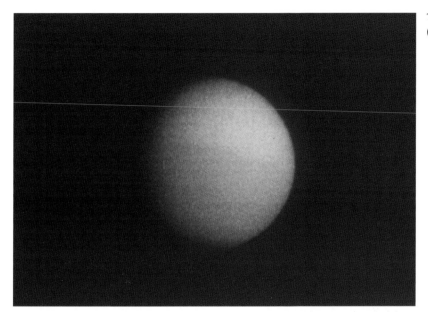

Titan as seen by Voyager cameras (NASA).

atmosphere containing methane and so we were expecting the Voyager images of it to be interesting. We thought that we would be able to peer down through that atmosphere to see the frigid surface, perhaps covered with some exotic compounds and organic molecules. It was, instead, a big disappointment. The first images that came back, when the spacecraft was still millions of miles away from Titan, just showed a little orange ball, fuzzy and blank.

When the spacecraft got close enough to get a really good image, what did it show? A big orange ball, fuzzy and blank. Sadly, Titan's atmosphere is opaque, smoggier than on the worst day in Los Angeles or Mexico City. Nothing could be seen of the mysterious surface because of this smog, and astronomers are still speculating about what that surface is like. Soon, in the year 2004, we have hope of answering the many questions that we have about it, as that is the year when the spacecraft Cassini reaches Saturn and its little probe named Huygens is dropped into Titan's atmosphere, to parachute down to the surface. Until then, we will have to wait a bit before making our detailed plans for exploring Titan on foot.

It's too cold on Titan for the smog to have been created by freeways and automobiles (even cars made in Sweden would have trouble starting at temperatures of −300 °F), so what causes it? Titan's atmosphere is continuously bombarded by the Sun's ultraviolet light and by particles from both the Sun and Saturn's ionosphere (the outer atmosphere of ions that have been captured within its giant magnetic field). Apparently, these result in chemical reactions in the nitrogen-rich atmosphere that are responsible for creating a wealth of different complex molecules. At least fifteen kinds of molecules have been detected so far, including such interesting organic compounds as acetylene and propane. You might think that with all these natural gasses present at least there wouldn't be any problem in keeping warm in your visiting spaceship, and you may have a point. But without free oxygen, these fuels are not combustible. When there, if you strike a match, not only will the atmosphere not go up in smoke, but neither will your match ignite.

The edge of Titan's atmosphere with its thick orange haze below thin blue cirrus-like clouds (NASA).

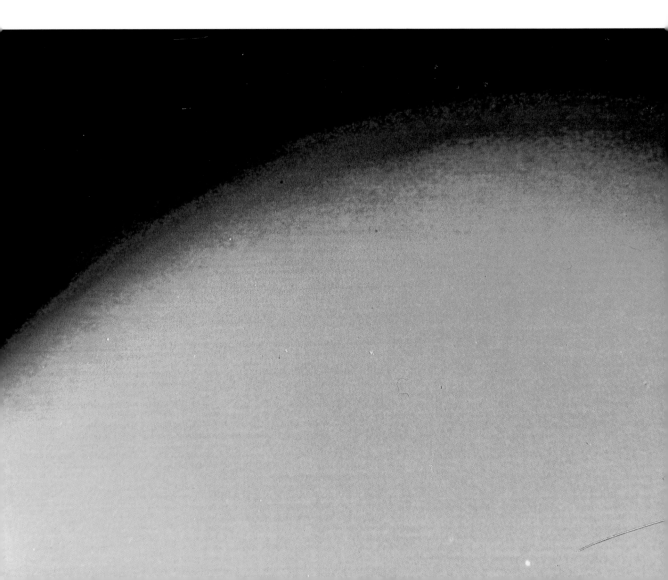

Oceans of ooze

It is still not known what is under the hazy atmosphere of Titan, but we have ideas. The particles that make up the smog probably slowly fall to the surface, where they may become a sticky, muddy layer of organic goo. Some astronomers estimate that this layer could be as much as several thousand feet thick. Hubble Space Telescope infrared images have detected what may be bright continents of highlands, perhaps with goo-covered mountains upon which methane raindrops fall, washing them clean of the worst of the foul-smelling tar.

But elsewhere on Titan, there may be lakes and seas. These could not be seas of water, but rather made of liquid ethane and methane. It is thought that, at the pressures and temperatures found at the surface, an accumulation of enough ethane (with a smattering of methane) might have built up over the lifetime of Titan to form oceans thousands of feet thick. We infer from the Space Telescope images that there isn't a global ocean, but there could be seas or lakes. And these bodies of water might be especially interesting to explore, as they may turn out to have important information in them about conditions on the early Earth, when it was first forming life from primitive organic compounds.

The late Carl Sagan, television star and skilled popularizer of astronomy, was also an excellent scientist, whose main passion was the search for life on other worlds. Titan intrigued him because of its plethora of organic materials. (It should be pointed out here that the term "organic" in this connection does not imply organisms of life. Chemists use the term to refer to any compounds that have carbon as a chief constituent and that include other elements that are common in life forms. To have life like us, it is necessary to have organic compounds, but to have organic compounds it is not necessary to have life.) Sagan and his colleagues proposed that Titan might be the key to our understanding of how life emerged on Earth, as the atmosphere of Earth billions of years ago was more like Titan's than like its present condition. Our atmosphere started out rich in hydrogen compounds, including methane, which were lost over time as the sun disassociated these molecules and they evaporated into space, leaving just carbon dioxide and water, which was steadily issuing from underground through volcanic activity. The molecules of oxygen upon which we are so dependent are a recent addition, placed in the atmosphere by plants.

Titan's night side, seen from behind, shows the atmosphere lit up by the Sun (NASA).

Most astronomers believe that the temperatures on the surface of Titan, in spite of a small greenhouse effect, are too cold for actual life to have evolved. But not everyone is convinced and one of the things that you will want to do while you're there is look around; are there any Titans stomping about through the muck? If any life forms do exist, they are probably microscopic in size and localized to small areas on the surface where warmth from volcanic-like activity might be available. One of your missions will be to search the surface for such likely oases and to bring back samples that might have little wiggly Titanoids in them.

[opposite] An imaginary view of Titan's surface, with its possible seas of liquid nitrogen or methane and outcroppings of ices and rock. The colors are exaggerated to emphasize the exotic and unknown nature of the surface.

The landing: splash, splot or clunk?

At this time we know too little about the surface of Titan to plan a landing. It might be that you will put down on an ocean, it might be a

swamp of organic goo, or it might be hard rock of some kind. Before you make your final plans it would be wise to check thoroughly the record sent back by the Cassini spacecraft to get a better idea of what to expect. Until that mission occurs, this chapter will have to remain rather tentative in its details. Plan to orbit the satellite for several days before the descent, gaining more information about the surface by using radar and infrared imaging, which can penetrate through the smog.

When you are finally ready for the visit to the surface, you will have picked out a best landing place. As in the case of the Apollo landings on the Moon, your choice will be a compromise between a safe, smooth place and an interesting one. Perhaps you will find a nice, smooth plateau next to the shores of an ethane lake and near to a warm volcanic area. The descent will probably be made with the help of a parachute, so you will also want to know something about the winds in Titan's atmosphere. We know little about them now, except that the temperature structure of the atmosphere tells us that there must be some strong global winds.

Once on the ground, it will be a matter of taking pictures and samples. Whether there is any walkable surface or not will decide whether you will exit from your spaceship. It may well be that you'll do everything remotely from inside, where it is warm and there is oxygen to breathe. But if you do emerge in your Titanic "space suit," your travels across the weird and wonderful landscape will be an amazing experience. Walking should not be difficult, as the gravity is sufficiently less than on Earth to make it easy to move about in spite of your heavy space suit. Your view may include lakes or oceans and even mountains. Perhaps you'll see a mountain on the horizon (if the smog allows) that rises up above the flat area where you have landed. And perhaps that mountain, if there is one, reaches high into the foul atmosphere. Maybe it will turn out to be a mountain climber's challenge, higher than Everest.

18

Climbing the cliff of Miranda

For centuries the Solar System beyond Saturn was an unexplored wilderness. The five major planets (Mercury, Venus, Mars, Jupiter, and Saturn) had been known since people first noticed their motion in the sky, which led to their being called the five "wandering stars." After Copernicus showed how they are arranged in distance from the Sun, it was assumed that Saturn, as the most distant planet, must be at the edge of the system. Superstitious people even believed that there was something significant in the number of planets and "7" became an important, almost magical number (there appeared to be seven moving objects in the sky: the sun, the moon and the five planets). Therefore it was a surprise and a shock when a new planet was discovered beyond Saturn.

William and Caroline

The first modern discovery of a planet was a complete accident. It was not, however, a mistake. The discoverer had set out to discover things; it's just that discovering a new planet was not one of the things that he anticipated doing. Although it's difficult to know historically what people at a certain time *didn't* think, we suspect that people of the eighteenth century didn't think that there were any more planets out there to be discovered.

William Herschel and his sister Caroline were living in the English city of Bath in the late 1700s. They had grown up in Hanover, Germany, but had moved to England for professional reasons. William made his living as a musician, playing a church organ, leading an orchestra and choir, and composing music for his students and instrumentalists to play. At night, however, he and his sister became astronomers. He developed a method for building large telescopes, larger and better than those in common use at the time. He eventually built the largest in the world, a remarkable fact considering that he was not a trained scientist and did not have a job as a scientist at a scientific institution. However, at the time there were very few "trained scientists" and virtually no scientific institutions. Much scientific work was done in the spare time available to men and women whose primary work was something quite different (consider

Benjamin Franklin, who made several important scientific discoveries, although he was by training a printer and by profession a statesman).

On the evening of March 13, 1781, Herschel was using a telescope about 6 inches in diameter, examining double stars, of which he had discovered a large number. Caroline was probably there, too, as they usually worked together. As they carefully mapped the sky that night, one of the stars he viewed looked odd to him. He noted that it appeared bigger than the rest. At first he guessed that it might be the disk of a comet, especially when he saw that its position changed from night to night. But its motion indicated that it must be very far away, farther than most visible comets, and that its orbit must be nearly a circular one around the Sun. Eventually it was realized that these facts indicated that the object was not a comet but a new planet, the first to be discovered since before written history began. It was a momentus discovery and Herschel was soon one of the world's most famous scientists. The English people were especially happy that the discovery was made in England and Herschel increased their delight by

Uranus from a distance, much as it looks from Earth through a large telescope (NASA).

naming the planet after the English King, George III. The King invited Herschel to the palace and awarded him a stipend so that he could continue his astronomical discoveries without having to give music lessons. William and Caroline Herschel proceeded to become the most prolific observational astronomers of their time, discovering large numbers of important objects and virtually creating the whole field of galactic astronomy. Caroline also made her name famous throughout Europe by being the discoverer of several bright comets.

But what of "George's Planet"? The rest of the world was not in favor of naming a planet after a living king rather than a classical god or goddess. At the suggestion of a German astronomer, J. Bode, eventually the name "Uranus" was adopted by the international community of scientists. It was based on the name of the ancient Greek god of the heavens.

Uranus turned out to be such a remote planet that very little was learned about it in the first two centuries of its study. At a distance of nearly 2 billion miles from the Sun, it was never more than a faint bluish blob seen in most telescopes from the ground. About all that was known was that it was large, about 4 times the Earth's diameter.

We now know a lot more. The Voyager 2 spacecraft visited Uranus in 1986, sending back lots of information about this mysterious object. It turns out to be a hydrogen-rich planet, with a deep watery interior and a rocky core. The Voyager photographs showed a nearly blank, ghostly blue surface, the strange color caused by small amounts of methane in its thick atmosphere, which is mostly hydrogen. Faint high clouds were just barely detected. Ten years later these clouds were seen again in images obtained with the Hubble Space Telescope, showing up as faint bands that go around the planet like lines of latitude.

But on Uranus these lines are at a weird angle. The planet's poles of rotation are nearly at right angles to the poles of its orbit, so that it turns on its side, with the result that it has extreme seasons. It's unique among the planets in this way. The other planets rotate roughly in the same direction as they revolve around the Sun, with most having inclinations (the angle between the pole of rotation and the pole of the orbit) less than 30°. Uranus' inclination is 98°, meaning that, technically, it even rotates backwards. Were you able to land near the equator on Uranus, your daylight would be about 17 hours long in the spring and fall. If you were near one of the poles, you would experience long, continuous darkness, as the pole would be pointed almost directly away from the Sun. In the

Uranus up close, as seen by
Voyager 2 (NASA).

Uranus up close, as seen by Voyager 2 (NASA).

summer, you would have many Earth years of sunlight. Of course the Uranian summer would still be neither warm nor bright, as the distant Sun would appear only about $\frac{1}{300}$ as bright as seen from Earth.

Shakespeare in the sky

This chapter's adventure will not land you on the surface of Uranus, for two reasons: first, the planet has no solid surface on which to land, and, second, one of its satellites makes a much more attractive goal. Like Saturn, Uranus has millions of moons, but most of them are tiny moonlets that revolve around the planet in narrow, dark rings. These rings are probably all shepherded by small moons, several of which are barely large enough to have been detected.

The five biggest of Uranus' moons are nicely sized and easily visible in telescopes. The largest two were discovered by William and Caroline Herschel six years after the discovery of the planet. Of the other three large satellites, two were found by William Lasell in 1851 and the fifth by Gerard Kuiper in 1948. Herschel named his two satellites after characters in Shakespeare, Titania and Oberon. This followed his practice of honoring his adopted country, England. Lassell continued this tradition with the satellite Ariel and also honored Alexander Pope by naming his other discovery Umbriel, after a character in one of Pope's books. A hundred years later, Kuiper returned to Shakespeare by naming his discovery Miranda.

The miracle of Miranda

As you might have guessed from the fact that it was so recently discovered, Miranda is the smallest of Uranus' major moons. It is only about 300 miles in diameter, hardly big enough for a country like the United Kingdom to be spread out on its entire surface. Its density is low, not more than half again as large as that of water ice, suggesting that it probably consists of a small rocky core covered by a thick mantle of ice. In this respect Miranda is like most of the moons of the outer planets, which are all heavily dominated by water ice in their interiors.

But Miranda does not look very much like the other moons when its surface is viewed up close. While Uranus' other large moons are nicely cratered, with their only unusual features being a few long narrow valleys that cross their dark, dirty and icy surfaces, the face of Miranda is covered with amazing features, unlike any found anywhere else.

Miranda's landscape shows only a few impact craters, indicating that its surface is not quite as ancient as those of its siblings. But its famous weird features are much larger than these craters. There are four strange formations on the surface. Two of them are immense oval things that look almost like giant football fields with rows of seating surrounding them. When first seen in images sent back from Voyager 2, each was called a "circus maximus" (this term refers to the large oval fields used for chariot races by the ancient Romans). Now, astronomers often use the less romantic term of "ovoids" for these peculiar features, which remain a miraculous and puzzling mystery. They occupy a large fraction of Miranda's surface, the long axis of each stretching over 125 miles. At first

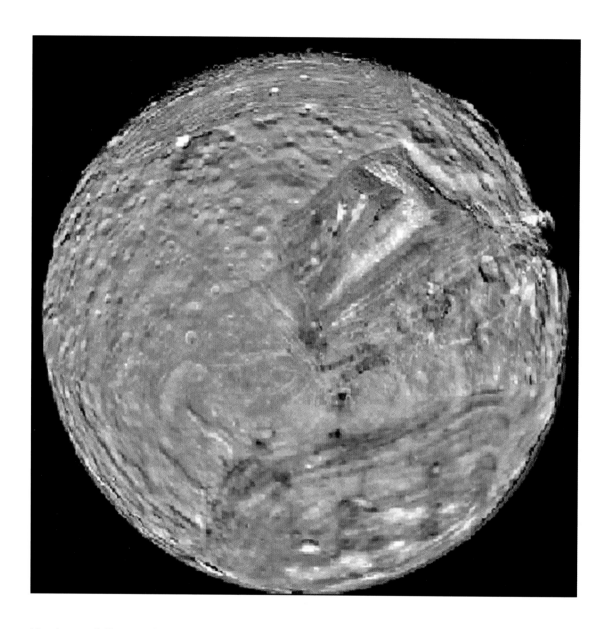

Miranda as seen by Voyager 2. The chevron is at the center right and the two circi maximi are at the bottom and just at the limb at the top of this image. The famous cliff is at the far right (NASA).

it was suggested that they might be huge chunks of Miranda's interior that had fallen back into the moon after it had been blown asunder by a catastrophic collision with another body. However, now it is thought that the most likely explanation is that the ovoids are the surface expression of huge convective bubbles that came up from below. Their official names are followed by the term "corona" on the basis of the idea that they may

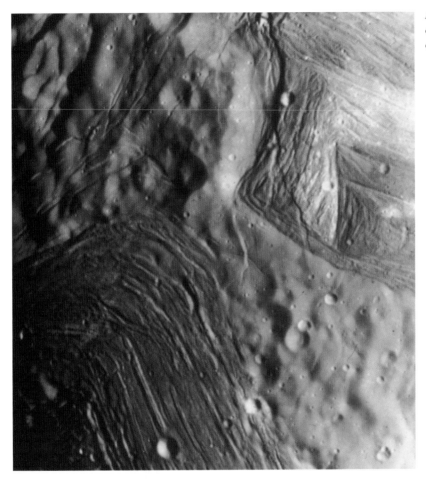

A close-up Voyager image of the chevron (right) and one of the circis maximi (left) (NASA).

be similar to the coronae found on Venus, though they look quite different in detail. Perhaps tidal heating by Uranus and its other moons warmed the icy interior enough to cause it to "boil" like water in a teakettle. Blocks of crust material may have been pushed and twisted and turned on their sides, cut by faults and covered by flows of icy lava.

The third weird formation is a thing originally called the "chevron," a rectangular feature with its interior made up of ridges arranged in a series of triangles. Its official name is Inverness Corona and it may have a similar origin to the two large ovoids. The chevron is about 75 miles across and at one corner there is a valley that leads out of it into an area where Miranda's most spectacular feature resides, its immense wall, the Solar System's highest cliff.

A Voyager close-up view of Miranda's miraculous cliff (NASA).

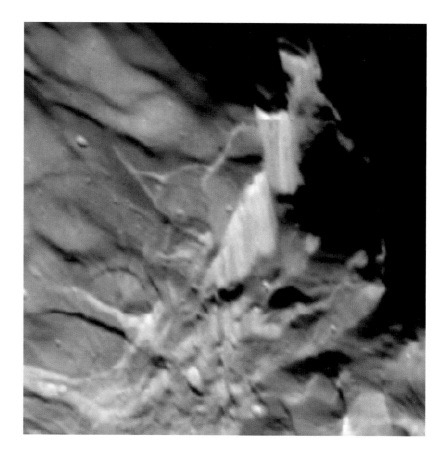

Higher than the Eigerwand

There are many famous cliffs on the Earth. The cliffs of Yosemite, Rainier's Willis Wall and the challenging face of the Eiger, known to mountaineers over the world as the Eigerwand, are only a few of the steep and difficult walls of rock that attract skilled and adventurous climbers to their smooth and dangerous faces. The Eigerwand is an especially famous example. It is the north face of the Eiger (the "ogre" in English), a high peak in the Bernese Oberland in central Switzerland. Rising a steep and fearsome mile above the scree at its base, the Eigerwand seldom sees the Sun. Ice and snow remain on its treacherous face even in summer. It was first climbed successfully (after several tragic fatal attempts) in 1938, when the famous mountaineer Heinrich Harrer and three companions made it to the top, taking several days for the ascent. They had to spend

The Eigerwand in Switzerland is one of Earth's famous cliffs, but is much smaller than Miranda's giant cliff (Swiss National Railways).

the nights perched at strategic places on the vertical rock before they could reach the summit. The first winter ascent was not made successfully until 1961, when four climbers attained the top after six days of climbing in bitter winter cold. Although they had achieved an impressive example of bravery and skill, they had to give back the gold medals awarded to them when it was revealed that they had not started from the very bottom, but had taken the mountain railway to get part way up at first. This railway is itself an amazing feature of the Alps; it winds its way through tunnels up into the heart of the adjacent peak, the Jungfrau, reaching a hotel and ski area 11,400 feet above sea level. But your goal is a peak even higher than the Eigerwand and there is no railway to help you with a head start.

The Eigerwand is about 5,000 feet from bottom to top. The cliff of Miranda is nearly 60,000 feet from base to summit. It takes a strong climber five or so days to climb the Eigerwand. How long will it take your climbing party to ascend this fearsomely taller cliff? It is difficult to know, but it is likely to be one of the great climbing achievements of the century (whichever century it turns out to be). It is probably primarily an ice cliff, though rocky material may also be there. It looks nearly perfectly smooth from Voyager's distance, though we don't know what the cliff face is like from up close. It looks nearly vertical from available images, though it may not turn out to be really that steep when we get there and look up at it.

Of course, it's going to be quite an accomplishment even to get there. Uranus is a remote planet. Even a supercharged spaceship (whatever that turns out to be) may take years to cover the nearly 2 billion miles distance from the Earth. The way that the outer Solar System stretches distances works very much against us when we plan travel out there. We are used to sending craft to Mars or Venus that take only months or a year or so to get there because the distances here in this part of the Solar System are measured in only tens of millions of miles. The distances between the planets grow, however, as we go out and, by the time we get to Uranus, we are talking about much larger distances. From Saturn to Uranus is twenty times the distance from Earth to Mars.

Once there, the plan will be to land on Miranda in a fairly smooth spot north of the chevron in a highlands area just south of the crater Prospero. You will probably plan to spend several days in this region, called Sicilia Regio, exploring this virgin world to learn as much as possible about its nature before departing the ship for the big adventure. It will be about twenty miles from the landing spot to the edge of the chevron, so some of the geologists aboard will probably want to make that excursion in the hope of solving the mystery of how it formed.

When you're ready, you'll load up your vehicle with food and climbing equipment and wend your way through the rugged valleys that lead to the base of Miranda's miracle cliff. The floors of these rugged valleys, which are collectively called Verona Rupes, are a jumbled mess. The distance is only about 80 miles, but it will probably take an Earth day or more to get there, as you must cross an unexplored landscape of icy hills. At least there should not be a problem of slippery ice. At Miranda's typical daytime temperatures ($-360\,°$F) the ice should provide a nice, hard surface.

What is a day on Miranda? Its orbital period is 33 hours, so technically that is the length of its day. It keeps one side always facing its parent planet, as does our Moon and most other satellites, so its rotation period and orbital period are the same. But in Miranda's case the situation is quite unusual. All of the Uranian satellites have orbits that are in the plane of the planet's equator. This means that Miranda's path in space is sometimes edgewise to the Sun and at other times face-on. When it's edgewise, it will have day and night every 33 hours. In the daytime the tiny Sun will share the sky with the dark side of Uranus and in the night the black, starlit sky will be dominated by the bright blue disk of the planet.

But at other seasons the story will be very different. When Uranus' pole is pointing almost directly towards the Sun, the moons will be circling the planet near the plane of the sky. Their poles will also be pointing towards the Sun and each hemisphere will be either in continuous darkness or continuous daylight. The cliff of Miranda is near its equator, so that, at this season, it will be witnessing a long dawn or dusk, with the sun arcing across the sky in a huge circle just above or below the horizon. It would be best to plan your visit to Uranus at a time when the cliff experiences daylight so that you will not need to carry extra batteries for your flashlights. Planning the visit for a time just before or after the spring or autumn solstice might be best, as then you would have some 16 hours of daylight and a 16 hour night, allowing you to have sunlight for climbing and darkness when you will need to rest.

Plan at least a day at the base of the cliff to make your plans for the ascent. You will want to survey the cliff carefully to figure out where you

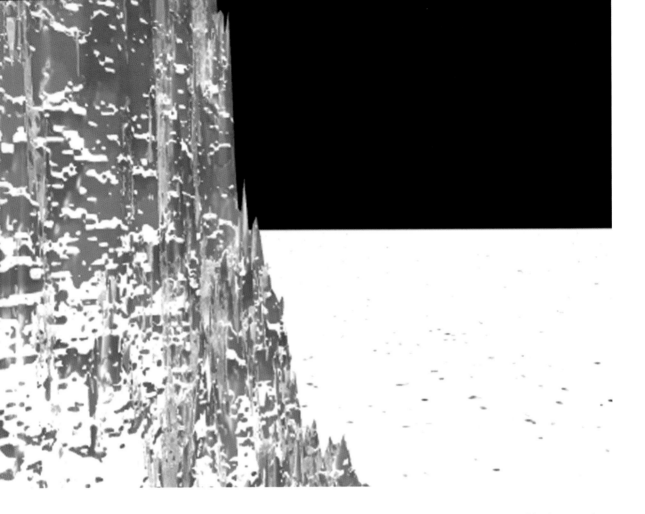

Miranda's cliff in an imaginary view.

might be able to bivouac for each rest period. You will also need to measure the steepness so as to be able to choose the right climbing equipment. If it turns out to be as nearly vertical as it looks from the Voyager pictures, you'll need to take specialized ice-climbing equipment and enough food for several weeks of climbing. But if it is not very steep, you may be able to ascend it in only a few days, using specially adapted crampons and ice axes, but no more technical equipment. The force of gravity that on Earth led to tragic falls on the Eigerwand will be much less on Miranda, which will help. But the distance from bottom to top is 12 miles, which is tremendously farther than the distance up the Earth's famous cliffs, and climbing it will be an incomparable achievement, especially if it does turn out to be as steep as it looks.

But even if it is not nearly vertical, the climb of Miranda's famous cliff will be a marvelous accomplishment in the annals of human adventure. You will have climbed the Solar System's highest cliff, much higher and more difficult than the Eigerwand (and, of course, higher than Everest).

19

The Yellowstone of the Solar System

The athletic feat of the previous chapter was phenomenal. Climbing a 60,000 foot-high cliff in frigid, airless weather is a major challenge, requiring tremendous amounts of skill and courage. For this chapter, therefore, we are planning a more relaxing trip. It has important scientific goals, but as an adventure it is more like a tourist excursion. You are going to visit an area not unlike Yellowstone National Park, but the distance from home is nearly 3 billion miles and the summer weather there is cold, with highs reaching about $-400\,°$F. It will take you a while to get there and you will want to be prepared for the cold, but otherwise it's an easy trip.

The backwards moon

William Lassell, the nineteenth-century businessman turned astronomer who had his own observatory near Liverpool in England, discovered many new things, but probably the most important of these was the large moon of the planet Neptune. It was 1846 and the planet itself had only been known for a few weeks when Lassell turned his telescope towards it. A tiny point of light near the planet's faint blue disk intrigued him and soon, after following it over a period of several nights, he realized that it was moving around the planet; it was clearly a satellite of Neptune.

Eventually this moon was named Triton after the mythological son of the Greek god Poseidon, ruler of the sea. Neptune was the Roman equivalent of the Greek Poseidon and the reason that his name was given to this planet was that it seemed at the time to be the lone denizen of the deep in our Solar System. As a god, Triton was well fitted out for his environment, the deep sea, as he was a merman, half-man and half-fish. But out on the plains of Greece he would have had trouble getting around in what, for him, would be an inhospitable environment. The moon Triton may seem at first to be similarly inhospitable to you when you reach it, mainly because of the extreme cold, but it has some spectacular geological features that will make its scenery worth the trip.

At the time of its discovery, astronomers saw right away that Triton had a strange characteristic. All of the planets and most of the satellites share the same direction of revolution in their orbits. If you were to look down on the Solar System from far above it in the north, the planets and moons would be revolving in a counterclockwise direction. It's as if the planetary police had erected a "One Way" sign in the Solar System and everybody was obeying it. This is not a mysterious phenomenon; astronomers explain it as the natural result of the way in which the Solar System evolved from a giant disk of gas and dust, which was revolving around the primitive Sun in that direction.

Neptune revolves and rotates in this same direction, but Triton is a maverick. Its orbit is backwards. It revolves around its planet in a clockwise direction (as seen from the north). The reason for this peculiar behavior is not known, though there have been speculations about it. Perhaps it once moved in the proper direction, but was disturbed early in the Solar System's history by an encounter with another body. Another possibility is that it was captured from the reservoir of icy bodies out beyond Neptune, which is called the Kuiper Belt. If this is the case, then Triton is a sibling of Pluto, the largest of the Kuiper Belt objects (the others out there are all very much smaller).

Triton's distance from Neptune is a little less than the Moon's distance from Earth, but because of Neptune's large mass (17 times the Earth's mass), its period is much shorter, only 5.9 Earth days, instead of a month. As it proceeds backwards around Neptune, it keeps the same face towards its planet, as does the Moon, so a person on its surface will experience 3 days of sunlight (feeble though it is) followed by three days of darkness. From the part of Triton that this expedition plans to visit, a region called Uhlanga, Neptune's deep blue disk will dominate the sky. In daytime it will be a large crescent largely in shadow, but as night falls it will appear increasingly fully illuminated until at Triton midnight there is a full Neptune. You will then be able to peer down into its misty blue, deep atmosphere and watch its white clouds rotate with the planet, crossing the visible disk in a mere 8 hours.

Deep freeze of the Solar System

Triton is a large satellite, 1,600 miles in diameter, which makes it almost as large as our Moon. It is less dense than the Moon, suggesting that it is not made up entirely of rock, but probably has an interior that includes about one-fourth water ice. The low density means a low surface gravity,

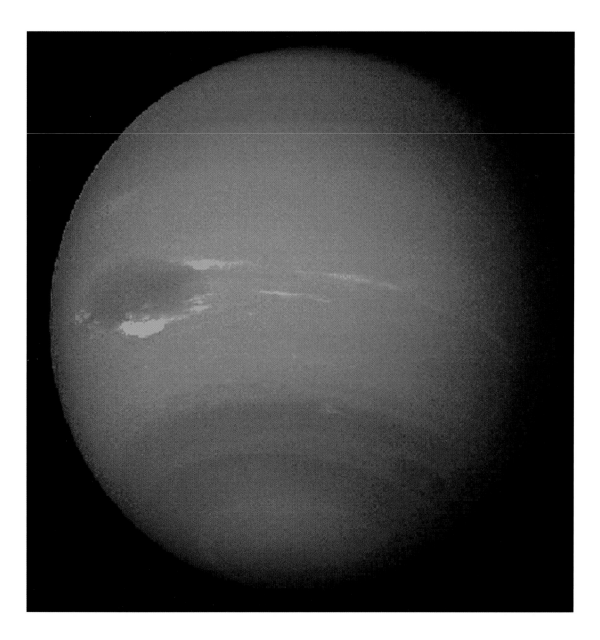

so that it will be fairly easy to walk around on the surface, even with a heavy space suit, much as the Apollo astronauts found in the case of the Moon. The surface will not look very much like the Moon's, however. From the way it reflects light, we know that Triton's surface is a bright white, not a dark blackish grey like our Moon's rocky surface.

Neptune as seen in a Voyager image (NASA).

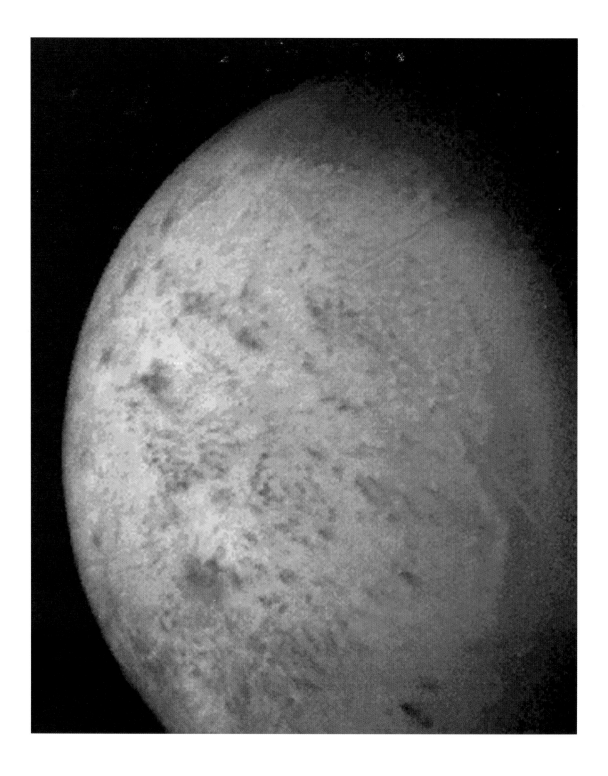

When astronomers analyze the light from Triton with a spectro-graph, they find further evidence that it is not like our Moon. The surface is made up of a mixture of various common and exotic ices. Normal water ice is abundant, as is "dry ice," frozen carbon dioxide. There is also evidence of frozen carbon monoxide, as well as methane and molecular nitrogen. These ices are made of the common elements of the outer Solar System, where the lighter elements dominate, as they do out in deep space where it is very cold. In fact, we have not found any place in the Solar System colder than the surface of Triton. Its temperature has been measured, even in mid-summer, to be about $-400\,°F$.

Triton is very cold and it is also nearly airless. But it does have a tiny atmosphere, made up, perhaps surprisingly, of the same gas that dominates our Earth's atmosphere, nitrogen. But the atmosphere is very thin, with a pressure of only about 40 millionths of that of our atmosphere. This means that you can't use it to breathe while you are there (anyway, it doesn't seem to have any oxygen, which you need), but it does permit Triton to have some interesting atmospheric effects, one of which is a thin haze. Another is the spectacular type of feature that you will witness as the goal of this expedition.

Frozen lakes and cantaloupes

The icy surface of Triton is beautiful and exotic. Astronomers had expected it to be a dead-looking, cratered wasteland. But when we first saw it in the Voyager pictures sent back during its 1989 encounter with Neptune, it presented an amazingly different face to us. There are very few impact craters there, indicating a youthful surface, the result of some kind of activity that renews it. Some areas are seen to be crossed by giant, curved and intersecting cracks, wide fault zones looking like those on Europa. Other areas show a weird, mottled appearance, resembling the surface of a cantaloupe (this was, naturally, called "cantaloupe terrain"). In other places there are giant frozen lakes, some as much as 50 miles across. Analysis indicates that these lakes are made up primarily of frozen water or a water–methane mixture. When we looked at the south polar region, which was facing us when Voyager was there, the long summer had already lasted 30 years. Near the edges of the polar cap we saw what look like melt regions.

[opposite] Triton's frozen surface as seen by Voyager 2 (NASA).

What could be the reason for all this geology on Triton? Somehow, here in the Solar System's coldest place, there has been activity – the ices have been melted and vaporized by some mysterious source of heat. One of the goals of the scientists on your expedition is to solve this mystery. We now think that Triton may be able to melt and vaporize some of its ices because of a special and unusual kind of Greenhouse Effect. Unlike that on Venus or the one beginning to take over on the Earth, Triton's may be a solid greenhouse, not an atmospheric one. It has been suggested that sunlight penetrates down into the transparent icy surface, where it is absorbed, heating the ice slightly. The warmed ice radiates away this energy in the form of infrared radiation, but the ices are not so transparent to the infrared, so that the energy is trapped there under the surface. This process may be enough of a source of heat to explain the active Triton geology, though there are other possibilities, too. Your expedition should be able to check these ideas and solve this mystery.

Geysers of the gods

When Voyager 2 reached Neptune, Triton's south pole had been experiencing a long summer. After 30 years of summer's feeble sunlight, the frozen nitrogen of the polar cap was partially evaporating into the atmosphere. The edge of the cap showed a change from bright ice to a darker surface, also icy, but grayer and more solid-looking. The ice of the cap looked mottled, with swirls and dips. The darker landscape at lower latitudes looked sharper, with hills and wrinkles and marked by several of the complex, long fault zones that slice across the satellite's surface.

In many ways the change from cap to its surroundings looks like that on the Earth or Mars, where the smooth, ice-covered terrain gives way to a sharp, rocky landscape. Between cap and the lower latitudes, the land is darker than elsewhere, forming a clear collar of countryside that experiences seasonal change from fresh ice cover in winter to bare surface in summer. It is a little bit analogous to the situation found in alpine mountains of the Earth, where the glacier-covered heights extend down to an intermediate zone, the smooth and flower-covered alpine "parks," below which are the rocky and tree-covered lowlands.

Not far from this polar cap edge are some remarkable features. When first seen in the images sent back by the Voyager spacecraft, they

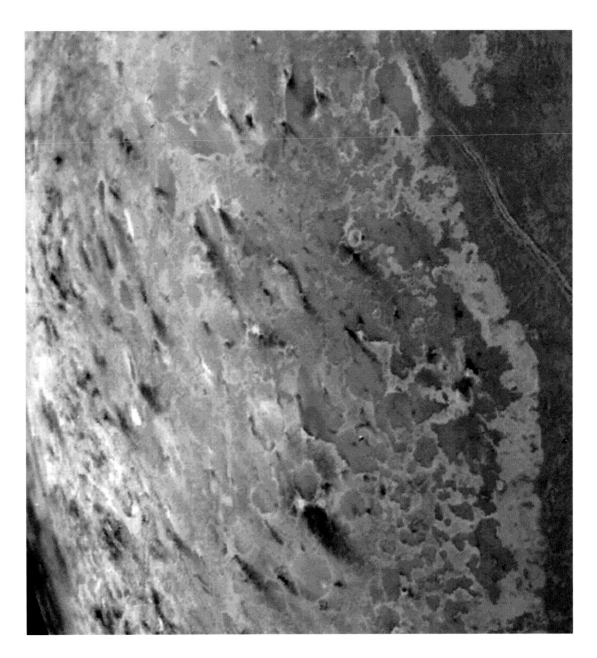

appeared to be just some vague, dark, elongated smudges. But, as more images arrived, these strange smudges provided a tremendous surprise. Between two pictures taken by the spacecraft 45 minutes apart, one of them appeared to have doubled in length. They were changing at a

Triton's south pole cap area. Geysers show up as bright spots with dark smudges to one side (NASA).

A terrestrial geyser in Yellowstone National Park (author photo).

remarkable rate. When the two images were viewed in stereo, scientists were amazed to see that the smudges were actually high above the surface, reaching heights of about 30,000 feet. It quickly dawned on the astounded observers that they were witnessing eruptions of giant geysers, vastly larger than even Yellowstone's mighty Old Faithful. The mild heating of the sub-surface polar ices on Triton appears to be enough to drive evaporated ices high into the thin air, where rapid jet streams then carry them north away from the pole.

Several of these gargantuan geysers were detected and three of them were sufficiently well-mapped to be given official names. Namazu was named after a mythical Japanese fish, Doro after a Nanay fishing god and Viviane after the mythical Welsh lady who was the object of Merlin's affections. Together these geysers occupy an area of some 30,000 square miles, ten times the area of Yellowstone National Park, one of Earth's feeble attempts to compete in the geyser business. It is to this geological wonderland that this chapter's adventure will take you.

A visit to Viviane

After the long trip from Earth to Neptune, it will no doubt be a great relief to go into orbit around Triton. At long last there will something to see up close and Triton's bright surface with its subtle colors and varied landscapes will be a pleasure to behold. Several days will be spent in mapping the surface, especially the area of the region named Uhlanga (a Zulu word), where Triton's geysers are found. A landing spot near Viviane will be chosen probably in a smooth-looking area at latitude $-33°$S, longitude $35°$W, between two nearly-circular features that are either frozen lake beds or ice-filled impact craters, about 30 miles across.

Once on the surface your exploring party will make its way towards the mysterious source of the Viviane geyser, some 10 miles away. Your vehicle will have a time of it, as the Voyager pictures suggest that the land is pretty rough in this region. Bouncing along over the icy surface may take you several hours, but it will surely not be boring. In this frozen land, the coldest place we know in the Solar System, there will be lots of exotic things to see.

When you finally reach your destination, what will you find there? We don't know. To propel its gasses so high into the atmosphere, Viviane must have a subsurface chamber where the evaporation occurs and an opening to the surface that focuses the gasses into an upward stream. Like Old Faithful, Viviane probably looks like a hole in the ground. We don't know yet whether, like Old Faithful, Viviane only spurts out its stuff at intervals. Instead, it may be constructed so that, throughout the Triton summer, it erupts a steady stream of evaporated ices. Probably a mixture of nitrogen and methane, these gasses turn dark at some point, possibly as they ascend into sunlight, which may be responsible for their dark color by converting the methane to organic molecules. This darkening process is also thought to explain some of the other blackish areas of Triton.

So what will you see? Probably, as you stand on the ice near Viviane's giant vent, you will see a stream of hazy vapor jetting out of the ground. As it rises high above you, it may turn darker, but will probably be visible as a grey mist against the black sky. If you are standing a little way off to the east, you should be able to see it rise almost straight up and then turn off to the right above you, as the jet stream winds catch it and carry it off to the north. Surely it will be a wondrous view as you stand there by this immense and exotic geyser, with bright and colorful ice all around you and the mysterious blue disk of Neptune off in the distant sky. And some of that giant plume is as high as, if not higher than, Everest.

20

All nine

This chapter is a little like the final few minutes of a fireworks display. After an evening of bright and colorful explosions, each separated by a discrete pause, the final event traditionally is a spectacular series of fireworks, set off in rapid succession. When this happens, everyone knows that the best is happening but it will soon be time to go home. This chapter is a similar sort of grand finale, in which an audacious and nearly-impossible task is set out for you. It is a burst of challenges and, if they are all met, it will surely be time to go home.

Peak bagging

Here on Earth, human adventurers have a seemingly unlimited ability to think up new goals to achieve and new records to break. It is important to us to be the first to climb the highest mountain, to reach the south pole on foot, to fly alone across the Atlantic, to bicycle across the Sahara, or to swim the English Channel. Even more locally there are firsts; mountain climbers who have made a first ascent, even of a minor peak, find it hard to not be pleased with themselves.

But the Earth is fast running out of "firsts." Hillary, Amundsen, Byrd, Earhart and their ilk have already claimed the best prizes, and less spectacular goals are fast being depleted. We have responded by inventing the compound achievement. It was too late for Reinhold Messner to be the first to climb Everest, so he became instead the first to climb all of the "eight-thousanders," the 14 mountains on Earth whose summits rise above 8,000 meters. On a more modest scale, outdoor clubs award pins and badges to those members who achieve some locally established compound achievement, such as rafting down a certain set of white-water rivers or summiting a particular list of local peaks.

A surprisingly large number of men and women have made it their goal to climb the highest mountains on every continent. And there are almost 100 people who have climbed to the highest point in every state of the United States, even including the rather ludicrous challenge of finding the highest point (or *any* high point) in some states, such as Florida. This chapter has a similar list: the highest points on all nine of the Solar System's planets.

As in the case of the 50-state challenge, the nine-planet list has its difficulties. Mountain tops are as scarce on Saturn as they are in Kansas. But that will not spoil this grand tour of planetary adventures; we will find a way to make each planetary visit an appropriate challenge.

The mountains of Mercury

By the time that humans are ready to meet this chapter's challenge, the problem of getting to the planets will be fairly straight-forward. And there probably will be a number of people who will have reached the lesser goal of traveling to all nine (though, by today's standards, that is a pretty special achievement to contemplate). Being quite close to the Earth, Mercury will be comparatively easy to reach.

The difficulty, therefore, is not access, but is of two other kinds. First, Mercury's surface is horribly hot. As discussed in Chapter 11, the surface in daytime reaches highs of over 800 °F. The members of your expedition will have to be prepared for the heat of the blazing, bright Sun, two and a half times larger in the sky than at Earth and almost seven times brighter.

The second problem is that we don't know where the highest mountain on Mercury is. Only a little over half of Mercury's surface has been mapped, so the highest point might be somewhere on the unexplored half. Even for the surface that we know about, the altitudes of only a restricted zone near the equator have been measured by radar and so we don't have a secure identification of the highest point. There are no high mountains like we find on Venus, Earth or Mars, so the highest point is not at all an obvious peak. Instead it is probably a feature, possibly a crater rim, which lies on a global rise, of which there are several. The radius of Mercury varies by as much as 5 miles, but this is mostly a departure from its being a perfect sphere (the Earth's shape is even less perfectly spherical). For example, there is a "bulge" near longitude 0° that has been blamed for the tidal capture of Mercury's rotation by the Sun.

But there are no obvious high mountain ranges. Mercury is a little like Iowa or Sussex; without a mountain range, finding the highest point becomes a matter of careful cartography. For Mercury, the necessary mapping is not yet complete. Until new spacecraft are sent there to map it more completely, we will use the Mariner photos to find a likely highest point and make tentative plans for it, being ready to revise these plans if necessary later on.

Mercury's largest known crater is a huge multi-ringed feature called Caloris Basin. It is somewhat like a lunar mare, but has not been filled with a smooth layer of lava as in the case of the Moon. Only half of Caloris Basin has been mapped so far, as half of it was in the shadow of night when Mariner 10 passed by each time (the spacecraft flew past Mercury three times, but each time the same part of the planet was in darkness). The crater is defined by three giant rings, achieving a diameter of some 800 miles. The outer ring forms a chain of hills and peaks called the Caloris Mountains, which reach to heights of a few thousand feet above the surrounding plains. It is among the southern Caloris Mountains that we choose to identify our goal, the highest point on the innermost planet.

The southern half of the Caloris Mountain chain is fairly rugged and includes some mountains that could be Mercury's highest. A peak at latitude 17°N, longitude 187.5°W is a good prospect. It has no name, so we'll just call it Sizzling Peak (caloris means "heat" in Latin and the basin was given this name because it lies near the hottest location on Mercury). There is a fairly smooth area just to the north of Sizzling Peak that would make a good landing spot. It is in the wrinkled inter-ring flats of Caloris Basin and there are a couple of low hills to traverse on your way south to the base of the peak. It is only about 10 miles to the valley that lies between Sizzling Peak and its neighbor to the east, a similarly high mountain. The climb will be a fairly easy one, though you will notice that Mercury's gravity is higher than the Moon's because of its large iron core, and so it will take more effort than, for example, Mt. Blanc, a similar mountain that is part of a lunar basin's ring, climbed in Chapter 6.

When you and your party of pioneers return to the Mercury lander, you'll be happy to cross off the first on your list. It will be time to start planning a trip to the next planet out.

Venus and the West Wall

Chapter 9 outlines a plan for reaching the top of Venus, the summit of Mt. Maxwell. But that voyage is a cautious one and does not qualify for a "climb." Instead of ascending this formidable mountain from its base, that expedition descends from above. It will be a much more remarkable achievement actually to climb Mt. Maxwell.

As long as you are going to the trouble of doing right by this mountain, you might as well make the climb a spectacular one by going up its

The Caloris Mountains, sketched from Mariner photos. "Sizzling Peak" is the one at the right of center.

formidable West Wall. Rising up uninterrupted from the high plain of Lakshmi, this steep wall is almost 20,000 feet high. The horizontal distance from a smooth landing place on the plain is nearly 100 miles. Your expedition will need to be a big one, made up of adventurers who will be prepared to spend weeks carrying supplies higher and higher up the steep slopes of this horribly hot and stifling planet.

We do not know what the surface on the West Wall is like. It may be smooth under foot or it could be a terrible jumble of rough rock. It may be a gentle, straight rise or it could be a series of nearly-impossible cliffs and ledges. It may be naked volcanic boulders or it could be rock covered by some mysterious kind of mineral "snow." Like the first people to attempt to conquer Everest, you will be challenging the unknown, but Maxwell is vastly more inhospitable in character and, of course, it is higher than Everest.

Earth and Mt. Everest

The future that this book addresses is one that combines tremendous technological advancement and a continuing thirst for adventure. I don't know just when this future will arrive, but I have faith that it will come. I also have faith that people here on Earth will develop a sense of reverence for their home planet and will, therefore, preserve some of its natural beauty and its geographical treasures. One such treasure is Mt. Everest.

At the end of the twentieth century the situation at Mt. Everest was grim but improving. Large numbers of climbers were swarming to its base, but the local government, with the support of the world's more responsible mountaineers, was limiting the number of people allowed to mount its frozen slopes. Several parties were organized to travel to the mountain, not to climb it, but to clean up the terrible piles of garbage left behind by less-responsible expeditions of the past. The stark beauty of the mountain and its aloofness had not yet been completely compromised by its fame. As time goes on, I hope that this trend towards restoring its pristine glory will continue.

If so, you may have to wait a while for your turn to climb Earth's highest. But the experience should be a good one for you. And even the most difficult and dangerous aspects of the climb will stand out in contrast to the much worse conditions experienced on the other planets. It will seem a relief to be here and that's how we should feel about our home.

Mars and Olympus

The high mountains of the Earth, with their glaciers and steep walls, offer their own special challenges (author photo).

Chapters 1 and 2 presented plans for climbing Mars' highest peaks. It wouldn't be much fun to follow the same route again, if someone has already done it by the time you're ready for Mars. So why not work out a new route? Perhaps no one will have ascended the steepest part of the basal wall of Olympus, which could become the most difficult part of a summiting strategy. Or perhaps you could add an exciting finale to the trip by descending deep into the summit crater after reaching the top.

Jupiter and the thunderheads

A mountain climber in the outer Solar System is faced with a dilemma.

The planets are giants and, by rights, ought to have giant mountains to climb. But they don't. They don't even have surfaces to walk on. The surfaces that we see from Earth are the tops of high clouds, beneath which are thick, poisonous atmospheres that get denser and denser as you go deeper until finally, in the twilight of great depths, the gasses change into the liquid hydrogen of a vast interior ocean.

Chapter 12 describes a trip to Jupiter's cloud layer and promotes a visit to its most spectacular and long-lasting storm, the Great Red Spot. That trip involves all kinds of difficulties, mostly caused by Jupiter's large mass: the strong pull of gravity, the dangerous radiation levels, the high winds, the lightning storms and the thick, unbreathable atmosphere. The highest clouds are made up of frozen ammonia crystals, stretched out in long east–west bands and swirling in bright rotating storms. The atmospheric pressure at the cloud tops is about one-tenth that at the Earth's surface and the temperature is a cool $-200\,°F$.

With no mountains to climb, the task at Jupiter will have to be analogous: finding and visiting the highest of its clouds. The storms among the clouds are constantly changing, with only the Great Red Spot having any semblance of permanence. Your first task will be to map the clouds at the time of your visit from orbit. Normally the highest clouds are found near the centers of active storms, where convection brings up great masses of gas from below. In a thunderstorm on Earth the center of the activity is often a towering white cloud called a thunderhead, which reaches far up into the stratosphere. We know that Jupiter's atmosphere is rife with lightning storms, so there should be plenty of thunderheads for you to map.

When you've located the (temporary) highest place among Jupiter's cloud tops, it will be your task to descend in your spacecraft to visit it. As explained in Chapter 12, the descent will not be simple because of Jupiter's high force of gravity and it will not be free of danger because of both the high radiation levels in the magnetosphere above the atmosphere and the high risk of being hit by lightning within the atmosphere. But dangers are the stuff of great adventures, and your trip should proceed with caution but not with timidity. And when you have reached the top of the highest cloud you can reflect on the fact that it lies much higher above the hot, murky and poorly defined "surface" of Jupiter than the top of Everest rises above Earth's sea level. The hydrogen of Jupiter's atmosphere remains gaseous to great depths, but at a few thousand miles below your spacecraft it is believed to be turned into liquid hydrogen by

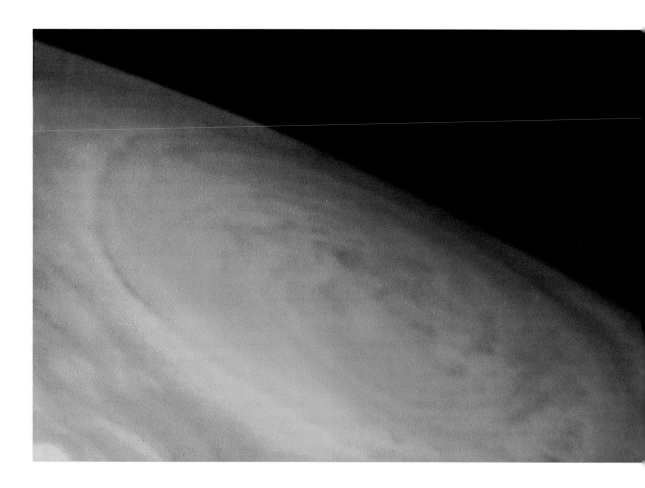

the high pressures found there. Thus, your visit to the tops of Jupiter's clouds will put you at an elevation of some 10 million feet above what we might call the "sea level" of this giant planet.

High clouds of Jupiter, seen in this oblique view of the Great Red Spot (NASA).

Saturn's big bulge

To reach the high point on Saturn requires merely going to its equator. Like Jupiter, Saturn has no solid surface. The atmosphere extends down to the point where the pressure is great enough that the gas, mostly hydrogen, becomes liquid. Saturn is almost as large as Jupiter with a diameter nearly ten times that of the Earth and it rotates very rapidly with a period of only 10 hours. This rapid rotation, combined with its large size and its largely fluid interior, gives Saturn an unusually large equatorial bulge. The diameter of Saturn measured through its equator is 10% greater than that measured through the poles. You can notice this difference clearly by examining any good photograph of the planet.

So visiting Saturn's highest point is simple. Just get there and descend into its atmosphere somewhere on its equator. You will be almost 2,000 miles farther from the center of the planet than if you were at an equivalent place in the atmosphere at a mid-latitude. By stretching meanings a little, you could claim that you had reached an altitude above the average "sea level" of Saturn of about 10 million feet.

Uranus, another one of "those"

Our imaginations are seldom disappointed in astronomy. The universe is so immense and full of so many amazing things that it usually stimulates our imaginations merely to gaze at an image of one of its wonders. Perhaps Uranus is an exception. It's another one of those giant planets without a nice, well-defined surface. But unlike Jupiter and Saturn, Uranus doesn't even have any pretty clouds to decorate the top of its atmosphere. When Voyager 2 reached it in 1986, the first images sent back to Earth showed a completely blank, ghostly blue disk – no markings at all. Only by careful and rather extreme image-processing were NASA scientists able to find even the slightest wisps of clouds, a couple of small elongated whitish things near the equator. More recently the Hubble Space Telescope has detected more clouds, but they are rare; the planet is mostly blank.

Of course, though Uranus may be disappointing to a mountain climber, it is pretty interesting otherwise. Its weird orientation, with its poles on its side, and its dark rings and remarkable moons make it a destination worth visiting. But once you have descended into its cold atmosphere and sought out some example of its high methane clouds, you might as well declare that you've reached the high point on this rather bland planet and return home.

Chasing Neptune's Scooter

Neptune, the next planet out, ought to look pretty much like Uranus. They are both in the far, cold reaches of the Solar System and they are nearly identical in size and mass. And both have hydrogen-rich atmos-

Neptune's white clouds, as seen by Voyager. The small triangular cloud was called "Scooter" (NASA Voyager photo).

pheres and mostly liquid interiors, with methane gas in the atmosphere giving them both a remarkable blue color when viewed through a telescope.

But Voyager 2 sent back some nice images of Neptune in 1989 that were quite beautiful. They showed a smooth, deep blue disk with dark spots and bands and some brilliant white clouds. The largest dark spot was called the Great Dark Spot and was found to be similar to the Great Red Spot on Jupiter. It was a giant storm region, with material rising from lower depths and rotating as a swirling equivalent of a cyclone. The white clouds were seen to be high in the atmosphere and most were drawn out into bands (parallel to the equator) by Neptune's rotation. One white spot, nicknamed the "Scooter," was a small triangular cloud that sped around the planet with higher velocity than the surroundings.

It might be comforting for people here on Earth who try to predict our weather to know that the weather on the other planets is also not so easy to understand. When the Hubble Space Telescope was turned towards Neptune in 1994, five years after the Voyager 2 encounter, the weather had changed drastically. The Great Dark Spot was gone and the white clouds were all different. There was no sign of the "Scooter" though new white clouds were found, especially near the poles.

So when you arrive at Neptune, the pattern will be new. There may be new dark storms and there surely will be some small white clouds, perhaps one with an unusually high speed that you can dub "Scooter." If so, that should be your goal, as it is probably the highest identifiable feature of Neptune. You will not only have the privilege of attaining some kind of a "first," but you can also contribute to the solution of the mystery of what causes the remarkable high winds of Neptune.

Pluto, the problem planet

The last in the planetary line-up is a strange little fellow. Pluto is not like any other planet. It is the smallest of the nine, less than half the size of modest little Mercury. It is made up mostly of ice, with little or no rock. Its surface is chiefly frozen nitrogen and it has a very thin atmosphere, made up of nitrogen and a little methane. There is evidence that the small amounts of methane may leave deposits of colored compounds here and

Pluto and its moon, Charon (NASA HST photo).

there on the surface, as the planet is measured to have a slightly reddish hue. Its poles are on its side; it rotates in about 6 days, but with its axis tilted like that of Uranus. The poles seem to have a bright deposit of ice on them, probably frozen methane.

An imaginary view of the frozen surface of Pluto. Its thin atmosphere provides a little color to the sky.

But Pluto is a problem. No spacecraft has visited it. It's the only planet so neglected. Our best maps come from reconstructions of surface brightness and color, which changed when its moon, Charon, passed in front of it many times during the interval 1987–91, when its orbit was lined up just right. The maps of its surface are not very detailed, but do show that there are bright and dark areas. It is thought that some of the bright areas are covered by frost that has been deposited out of the methane and nitrogen atmosphere. Also, some may be relatively young impact basins, bright because of the way in which the impact uncovered fresh ice from below the surface. Dark areas may be older terrain that has been "stained" in time by the conversion of methane ice into more complex, darkish, reddish compounds.

But where is the highest point on Pluto? We have no idea. The polar regions seem to have bright polar caps, but nothing is known about altitudes there or anywhere else. Your task will be to orbit the planet first, mapping its surface (if a robotic spacecraft hasn't already done so by then) and determining its topography. Then, having located the right place, you can land on its desolate and frigid landscape and make an attempt to climb its highest mountain. Although the temperature will be horribly low ($-380\,°F$), at least you will be able to take advantage of the light gravity. Pluto's small mass means that the surface gravity is only about one-tenth that of the Earth. So, even with your bulky space suit on, you can probably bound up the slopes of whatever Pluto provides as a mountain, which is very unlikely to be higher than Everest.

Index